Lars Hofer

Rare hadronic B decays

Lars Hofer

Rare hadronic B decays

in the MSSM and in other models of new physics

Südwestdeutscher Verlag für
Hochschulschriften

Imprint
Any brand names and product names mentioned in this book are subject to trademark, brand or patent protection and are trademarks or registered trademarks of their respective holders. The use of brand names, product names, common names, trade names, product descriptions etc. even without a particular marking in this work is in no way to be construed to mean that such names may be regarded as unrestricted in respect of trademark and brand protection legislation and could thus be used by anyone.

Publisher:
Südwestdeutscher Verlag für Hochschulschriften
is a trademark of
Dodo Books Indian Ocean Ltd., member of the OmniScriptum S.R.L Publishing group
str. A.Russo 15, of. 61, Chisinau-2068, Republic of Moldova Europe
Printed at: see last page
ISBN: 978-3-8381-2473-5

Zugl. / Approved by: Karlsruhe, KIT, Diss., 2011

ABSTRACT

In this thesis we study several aspects of new physics in rare hadronic B decays. First we consider the Minimal Supersymmetric Standard Model (MSSM) with Minimal Flavour Violation (MFV). Here, interesting effects arise for large values of $\tan\beta$ due to the enhancement of down-quark self-energies. These effects are well-studied within the decoupling limit, i.e. in the limit of supersymmetric masses far above the electroweak scale. In this thesis we address this topic in a framework that goes beyond this limit: We derive several resummation formulae for arbitrary values of the supersymmetric mass parameters and clarify their dependence on the renormalisation scheme. Furthermore, we study $\tan\beta$-enhanced corrections to couplings involving genuine supersymmetric particles. This cannot be done consistently in the decoupling limit with these particles being integrated out. We demonstrate that $\tan\beta$-enhanced corrections induce flavour-changing gluino couplings which have a large impact on the Wilson coefficient C_{8g} of the chromomagnetic operator. To illustrate the phenomenological consequences of the new gluino contribution to C_{8g}, we discuss its effect on the mixing-induced CP asymmetry in the decay $B^0 \to \phi K_S$. Our resummed $\tan\beta$-enhanced effects are cast into effective Feynman rules permitting an easy implementation in automatic calculations.

In the second part of the thesis we investigate the possibilities of probing new physics in the electroweak penguin sector via rare hadronic B decays. This kind of new physics is suggested by the measurement of $\Delta A_{\rm CP} \equiv A_{\rm CP}(B^- \to K^-\pi^0) - A_{\rm CP}(\bar{B}^0 \to K^-\pi^+) \stackrel{\rm exp.}{=} (14.8 \pm 2.8)\%$ which approximately vanishes in the Standard Model. After performing an updated analysis of $B \to K\pi$ using the framework of QCD factorisation, we conclude that, in order to clarify the picture, one should also consider other decay channels which are sensitive to the electroweak penguin sector. Apart from the analogous $B \to K\rho, K^*\pi, K^*\rho$ decays, we propose to study the purely isospin-violating decays $B_s \to \phi\pi, \phi\rho$ which are dominated by the electroweak penguin topology. In a model-independent analysis we study a potential enhancement of $B_s \to \phi\pi, \phi\rho$ in light of a χ^2-fit of the model parameters to $B \to K\pi$ data and with respect to constraints from other hadronic B decays. We find that in most scenarios an enhancement by an order of magnitude is possible. Given this situation, we study two concrete scenarios: Models with a modified flavour-changing Z-coupling and models with a flavour-changing Z' coupling. In such scenarios also constraints from semileptonic B decays and from B_s-$\bar{B}_s$ mixing arise.

The results of this thesis are published to some extent in Refs. [1, 2].

CONTENTS

1 Introduction **1**

1.1 Rare B decays in the MSSM with Minimal Flavour Violation 3

1.2 Probing new physics in electroweak penguins via hadronic B decays 6

2 General framework for the analysis of hadronic B decays **9**

2.1 The effective $\Delta B = \Delta S = 1$ Hamiltonian 10

2.1.1 Definition and construction 10

2.1.2 Renormalisation group evolution 11

2.2 QCD factorisation 13

2.2.1 The factorisation formula 14

2.2.2 Properties and limitations of QCDF 18

2.2.3 Input parameters 19

I Rare B decays in the MSSM with Minimal Flavour Violation 23

3 The minimally flavour violating MSSM **25**

3.1 Construction of the MSSM 25

3.2 Minimal Flavour Violation 28

3.2.1 Symmetry-based definition of MFV 28

3.2.2 Naive MFV 31

3.2.3 Low-scale structure of the MSSM with MFV 32

4 Resummation of $\tan\beta$ - enhanced loop corrections beyond the decoupling limit **35**

4.1 Effective theory approach for $M_{\text{SUSY}} \gg v, M_{A^0,H^0,H^\pm}$ 35

4.2 Diagrammatic resummation 38

4.2.1 $\tan\beta$-enhancement in $\mathcal{M}$ 39

4.2.2 $\tan\beta$-enhancement in $\mathcal{L}_{\text{ct}}$ 40

4.3 The flavour-conserving case 41

4.3.1 Flavour-conserving $\tan\beta$-enhanced self-energies 41

4.3.2 Renormalisation of the Yukawa coupling 43

4.3.3 Scheme dependence of the resummation formula 45

4.3.4 Self-energies in internal quark lines . . . 48
4.4 Flavour mixing I: External leg corrections . . . 49
4.4.1 Flavour-changing $\tan\beta$-enhanced self-energies . . . 49
4.4.2 Flavour-changing self-energies in external quark legs . . . 50
4.4.3 QCD corrections to flavour-changing self-energies . . . 52
4.4.4 Renormalisation of the CKM matrix . . . 54
4.5 Flavour mixing II: Flavour-changing wave-function counterterms . . . 56
4.5.1 Flavour-changing wave-function renormalisation . . . 57
4.5.2 Formulation of effective Feynman rules . . . 60

5 Phenomenology: Rare non-leptonic B decays beyond the decoupling limit 63
5.1 Effective FCNC couplings . . . 63
5.2 Gluino contributions to the effective $\Delta B = 1$ Hamiltonian . . . 65
5.3 The mixing-induced CP asymmetry in $\bar{B}^0 \to \phi K_s$ . . . 69

II Probing new physics in electroweak penguins via hadronic B decays 73

6 Isospin violation in $B \to K\pi$ decays 75
6.1 Isospin decomposition of the amplitudes . . . 75
6.2 Topological parametrisation . . . 77
6.3 Current status of isospin violation in $B \to K\pi$ . . . 80
6.3.1 Direct CP asymmetries . . . 81
6.3.2 Branching fractions . . . 83
6.3.3 Mixing-induced CP violation . . . 84
6.4 The decays $B \to K\rho, K^*\pi, K^*\rho$ . . . 85

7 The purely isospin violating decays $B_s \to \phi\pi, \phi\rho$ 87

8 Model independent analysis 91
8.1 Modified EW penguin coefficients . . . 91
8.2 Fit to $B \to K\pi$ data . . . 94
8.2.1 The *R*fit method . . . 94
8.2.2 Results . . . 96
8.3 Consequences for $B \to \phi\pi, \phi\rho$ . . . 98

9 Survey of viable NP models 101
9.1 Constraints from $\bar{B} \to X_s\ell^+\ell^-$, $\bar{B} \to K^*\ell^+\ell^-$ and B_s-$\bar{B}_s$ mixing . . . 101
9.2 Flavour-changing Z-boson coupling . . . 103
9.2.1 Effective Hamiltonian . . . 104

9.2.2 Results 105
9.3 Models with an additional $U(1)$ gauge symmetry 106
9.3.1 Effective Hamiltonian 107
9.3.2 Results 109

10 Conclusions **111**

A Appendix **115**
A.1 Sparticle mixing 115
A.1.1 Squark mixing 115
A.1.2 Chargino mixing 116
A.1.3 Neutralino mixing 117
A.2 Loop functions 118
A.3 Feynman rules for large $\tan\beta$ 119
A.4 Gluino contributions to the $\Delta B = 1 = \Delta S = 1$ Hamiltonian 120
A.5 QCDF results for $B \to K\rho, K^*\pi, K^*\rho$ 122

1. Introduction

The Standard Model of particle physics (SM) describes the electromagnetic, weak and strong forces among the known elementary particles (quarks, leptons and gauge bosons) with very high precision. Up to energies currently available at accelerator experiments, no significant discrepancies between theory and experiment have been found yet. Despite this tremendous success, most physicists regard the SM only as an effective theory which has to be replaced by a more fundamental one above the TeV scale. Experimental evidence for new physics (NP) beyond the SM are the by now established, non-vanishing neutrino masses as well as the strong hints for the existence of non-baryonic dark matter. Moreover, the SM suffers from the so-called hierarchy-problem: The mass of the Higgs boson which should be of the same order of magnitude as the electroweak vacuum expectation value (vev) $v \approx 174\,\mathrm{GeV}$ is not protected by a symmetry in the SM. As a consequence it is pushed via quantum corrections to the scale where NP enters, for example to the Planck scale $M_{\mathrm{Pl}} \sim 10^{19}\,\mathrm{GeV}$ at which gravity has to be incorporated into the theory. To stabilise the Higgs mass nevertheless at the electroweak scale, one has to fine-tune the parameters in the Lagrangian to an extent regarded as unnatural.

In the SM, flavour changing neutral currents (FCNCs) are heavily suppressed. Their amplitudes involve small elements of the quark mixing (CKM) matrix and in addition also a small loop factor because the SM does not provide a tree-level FCNC coupling. Furthermore, in most cases their size is even further reduced, for example by destructive interference of the contributing Feynman diagrams (Glashow-Iliopolus-Maiani (GIM) mechanism) or by the appearance of small ratios of quark masses (helicity-suppression). The origin of all these suppression effects is the very special structure of the Yukawa matrices which are in the SM the only source breaking the global $U(3)_Q \times U(3)_u \times U(3)_d$ family symmetry of the gauge sector. Being thus an accidental property of the SM, the suppression of FCNCs is absent in generic NP extensions. For this reason, FCNC processes are an ideal place to look for NP at the TeV-scale, complementary to the direct searches at LHC. Nowadays many NP scenarios are already highly constrained by data from flavour physics and once, new particles are found in high-p_T experiments, the flavour physics experiments will help to determine their properties (e.g. their couplings).

In this thesis we study rare B decays mediated at the quark-level by a $b \to s$ transition, with special focus on non-leptonic decays into two mesons. The $b \to s$ transition occurs at the low scale $\mu \sim m_b$ and is described by point-like interaction operators Q_i of an effective Hamiltonian

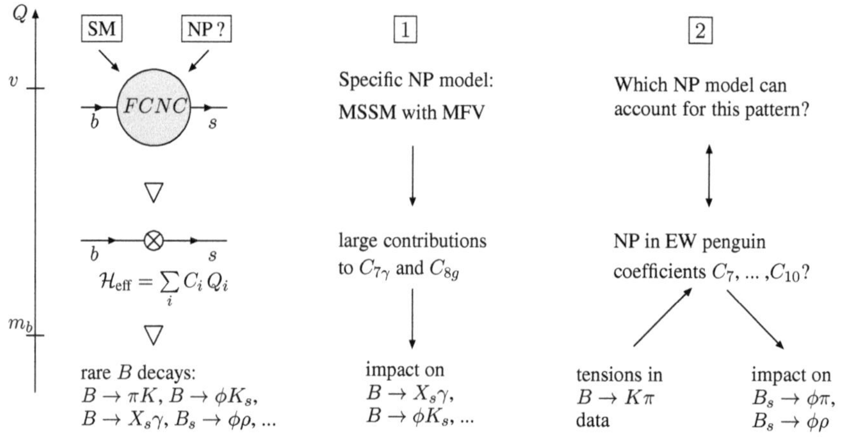

Figure 1.1: Two strategies to explore new physics in FCNCs.

$\mathcal{H}_{\text{eff}}$. It is, however, sensitive to high-scale physics at $\mu \gtrsim v$ because the corresponding couplings C_i are induced by loops of virtual heavy particles, to which, beyond the SM W-boson and top-quark, also new particles might contribute. From the universal effective Hamiltonian $\mathcal{H}_{\text{eff}}$ one can then calculate branching ratios, CP asymmetries and other B-decay observables (left pictogram in fig. 1.1). In this context hadronic B decays into two mesons are, on the one hand, very attractive since they offer lots of decay channels allowing for over-constraining measurements of the couplings C_i. On the other hand, non-leptonic B decays suffer from large uncertainties caused by QCD effects which hide the information on the coefficients C_i. Whereas one can take control of large QCD effects above the scale $\mu \sim m_b$ by a renormalisation group improved treatment, the low-energy QCD effects inducing the confinement of quarks into hadrons at the scale $\mu \sim \Lambda_{\text{QCD}}$ still pose a problem. Methods developed so far rely on flavour symmetries of QCD or on the factorisation properties of low-energy QCD dynamics (QCDF) [3–5]. Unfortunately, none of the two is able to predict the decay amplitudes with the required precision. The former is applicable only to a handful of decays while the latter, which implies an expansion of the amplitudes in Λ_{QCD}/m_b, receives important contributions from a number of subleading terms which can only be estimated. Throughout this thesis we will use the QCDF framework. The basic idea and the main features as well as our conventions are discussed in Chapter 2.

There are two strategies which can be pursued in order to explore NP in FCNCs (see figure 1.1). The first is to choose a specific, well-motivated NP model, calculate the FCNC couplings C_i in terms of the free model parameters and analyse whether large contributions to certain B

decays are predicted. The confrontation with experimental data will then impose constraints on the model parameters. In the first part of the thesis, this approach is applied to the Minimal Supersymmetric Standard Model (MSSM) with Minimal Flavour Violation (MFV), one of the most popular and widely studied NP models proposed so far. Even though it is constructed in order to avoid dangerously large FCNCs, the MFV scenario still permits large effects on some FCNC couplings, especially on $C_{7\gamma}$ and C_{8g}, the coefficients of the magnetic and chromomagnetic operators. We will identify and consistently resum enhanced contributions to these coefficients and study their impact on rare B decays like $B \to X_s\gamma$ and $B \to \phi K_s$.

In contrast to this "top-down" approach, the second strategy takes existing tensions in the experimental data on rare B decays as starting point of the analysis. The universal structure of the low-energy effective Hamiltonian $\mathcal{H}_{\rm eff}$ then allows to perform a model-independent study with the objective to explore whether additional contributions to certain couplings C_i could account for these tensions. Taking the resulting modified coefficients C_i as basis, one can then study their impact on other decays as well as search for a NP model which provides appropriate contributions to these couplings. This "bottom-up" approach is pursued in the second part of the thesis. Interpreting tensions in the $B \to K\pi$ data ("$B \to K\pi$ puzzle") as hints on NP in the electroweak-penguin couplings $C_7, ..., C_{10}$, we investigate the consequences on the purely isospin-violating $B_s \to \phi\pi$ and $B_s \to \phi\rho$ decays and discuss several viable NP scenarios.

The remainder of this Chapter is dedicated to a more detailed description of the two projects. The status quo of the corresponding research fields is stated and the new aspects contributed by this work are pointed out.

1.1. Rare B decays in the MSSM with Minimal Flavour Violation

As already explained in the outset, the hierarchy problem encountered with the SM calls for a symmetry which protects the mass of the Higgs boson. In supersymmetric theories (SUSY), bosons and fermions are related to each other and arranged in pairs of superpartners. The symmetry ensures that quantum corrections caused by such a pair of partner particles exactly cancel each other and thus it provides a natural solution to the hierarchy problem. Furthermore, SUSY offers a candidate for Dark Matter and it helps to unify the electromagnetic, weak and strong interactions at a high scale [6]. Among the supersymmetric extensions of the SM the one with the fewest number of particles and interactions is called Minimal Supersymmetric Standardmodel (MSSM).

An exact realisation of SUSY in nature would require the partner particles to have equal masses. Therefore the fact that no partner of a SM particle has been detected up to the present implies that SUSY has to be broken. Because of ignorance of the correct mechanism, SUSY breaking is usually parametrised by writing terms into the Lagrangian which violate SUSY explicitly. These terms have an a priori arbitrary flavour structure which is however constrained from low-energy data on quark and lepton flavour transitions. The resulting constraints are quite tight and suggest that the SUSY breaking terms obey a pattern of Minimal Flavour Violation (MFV) in the sense that they do not introduce additional sources of flavour violation compared to the SM. A consistent, symmetry-based definition of MFV has been given in Ref. [7]. For our studies we adopt a simplified version of MFV which we call naive MFV and which includes the widely studied CMSSM (see e.g. Refs. [8] for recent studies) and mSUGRA [9] models.

Even though the MSSM with MFV mimics the flavour structure of the SM with its strong suppression of FCNC transitions, there can still arise big effects in some flavour observables. These effects are caused by a parametric enhancement originating from the Higgs sector. The MSSM contains two Higgs doublets H_u and H_d coupling to up- and down-type quark fields, respectively. The neutral components of these Higgs doublets acquire vacuum expectation values (vevs) v_u and v_d with the sum $v_u^2+v_d^2$ being fixed to $v^2 \approx (174\text{ GeV})^2$ and the ratio $\tan\beta \equiv v_u/v_d$ remaining as a free parameter. Large values of $\tan\beta$ (~ 50) are theoretically motivated by bottom-top Yukawa unification, which occurs in SO(10) GUT models with minimal Yukawa sector, and phenomenologically preferred by the anomalous magnetic moment of the muon [10]. Since a large value of $\tan\beta$ corresponds to $v_d \ll v_u$, it leads to enhanced corrections in Feynman amplitudes where the tree-level contribution is suppressed by the small vev v_d but the loop-correction involves v_u instead. In such cases the ratio of one-loop to tree-level contribution receives an enhancement-factor $\tan\beta$ which may lift the loop-suppression rendering the ratio of order $\mathcal{O}(1)$ [11].

These $\tan\beta$-enhanced loop-corrections lead to a plethora of phenomenological consequences. They modify the relation between the down-type Yukawa couplings y_{d_i} and the quark masses m_{d_i} [11] and renormalise the elements of the CKM matrix [12]. These effects modify the neutral [13] and charged Higgs [14] phenomenology, with large impact on $B^+ \to \tau^+\nu$ [15] and $B^+ \to D\tau^+\nu$ [16]. Furthermore, $\tan\beta$-enhanced loop-corrections induce FCNC couplings of the neutral Higgs bosons to down quarks [17] with most spectacular effects in $B_{d,s} \to \ell^+\ell^-$. A priori one could have hoped to find an enhancement by several orders of magnitude compared to the SM value in this mode, caused by six powers of $\tan\beta$ [18]. The current upper limit on $\text{Br}(B_s \to \mu^+\mu^-)$ from Tevatron [19] puts severe constraints on the Higgs sector of the large-$\tan\beta$ MSSM and at the same time renders analogous neutral Higgs effects in B_s-$\bar{B}_s$ mixing invisible [20]. A different type of $\tan\beta$-enhanced corrections occurs in Higgs couplings of the right-handed top-quark field, which are suppressed by a factor of $\cot\beta$ at tree level. Supersymmetric vertex corrections can lift the $\cot\beta$ suppression, so that the one-loop correction competes with the tree-level coupling. This, for example, affects the $\bar{t}_R s_L H^+$ coupling entering the charged-Higgs loop in $b \to s\gamma$ [21,22].

In order not to spoil the perturbative expansion a special treatment is required for all these $\tan\beta$-enhanced corrections to resum them to all orders. There are two possible ways to deal with them.

The first method is to consider an effective theory with the SUSY particles integrated out keeping only Higgs fields and SM particles. This approach, which has been used in nearly all the phenomenological studies mentioned above, is valid for $M_{\text{SUSY}} \gg v,\ M_{A^0,H^0,H^\pm}$, i.e. within the decoupling limit; it can be extended beyond using an iterative method [23, 24] which converges if the magnitude of the resummed corrections is numerically smaller than the tree-level value. The second possibility is to perform a diagrammatic, analytic resummation in the full MSSM without assuming any hierarchy between M_{SUSY}, $M_{A^0,H^0,H^\pm}$ and v. This method has been developed in Ref. [14] to determine the modified relation between the quark Yukawa couplings y_{d_i} and m_{d_i} beyond the decoupling limit and it has been applied in Ref. [25] to the lepton sector in an analysis of the muon anomalous magnetic moment. In this thesis we extent the diagrammatic resummation method to the case of flavour-changing interactions permitting thereby its application to the phenomenological effects listed above, like the CKM renormalisation and the FCNC neutral Higgs couplings. Going beyond the decoupling limit in this way is desirable for the following reasons: On the one hand, $M_{\text{SUSY}} \sim v$ is natural since an unnatural fine-tuning in the Higgs potential is needed to achieve $M_{\text{SUSY}} \gg M_{A^0}, M_{H^0}, M_{H^+}$ [26] and since after all the widely-studied scenarios with neutralino LSP involve several SUSY particles with masses around and below v. On the other hand, $\tan\beta$-enhanced effects in couplings involving SUSY particles like gluinos and neutralinos cannot be studied in an effective theory with these particles integrated out. In this thesis we fill this gap and derive moreover analytic formulae for all the other effects where up to now only results valid in the decoupling limit or iterative methods have been available (see Tab. 1.1). The only exception are proper vertex corrections like in the $\bar{u}_{i,R}\, d_{j,L}\, H^+$ coupling which are relevant only for a handful of processes. In such a case a treatment beyond the decoupling limit requires a full NLO calculation for the particular process.

Our discussion of $\tan\beta$-enhanced effects is organised as follows: In Chapter 3 we give a brief introduction into the MSSM, discuss in detail the hypothesis of MFV and define our simplified framework of naive MFV. Chapter 4 is then dedicated to the development of our diagrammatic resummation technique. After a brief review of the effective theory approach in Section 4.1 pointing out its shortcomings, we describe the basic idea of the diagrammatic method in Section 4.2. It is applied to the flavour conserving case in Section 4.3, where we extend the results of Ref. [14] by a discussion of the renormalisation scheme dependence of the resulting resummation formula. In the flavour changing case $\tan\beta$-enhanced corrections are induced by self-energy insertions into external quark legs. Here we present two different approaches: In Section 4.4 the corresponding diagrams are treated as one-particle irreducible. The inclusion of $\tan\beta$-enhanced corrections into a LO calculation requires then the identification and calculation of certain NLO diagrams. This method has already been worked out in the diploma thesis [27] and it has been applied there to two particular processes, namely $B \to X_s\gamma$ and B_s-$\bar{B}_s$-mixing. Here we embed it into the systematic framework set up in Section 4.2. Furthermore, Section 4.4.3 delivers a piece

effect	decoupling limit	beyond
modified relation $y_{d_i} \leftrightarrow m_{d_i}$	Hall,Rattazzi,Sarid [11]; Carena,Olechowski, Pokorski,Wagner [11]	Carena,Garcia, Nierste,Wagner [14]; $\boxed{4.3.3}$
corrections to CKM matrix	Blazek,Raby,Pokorski [12]	Buras,Chankowski, Rosiek,Slawianowska [23]; $\boxed{4.4.4}$
enhanced FCNCs $d_i\, d_j\, H^0/A^0$	Hamzaoui,Pospelov,Toharia [17]; Babu,Kolda [18]; Buras,Chankowski,Rosiek, Slawianowska [20]	Buras,Chankowski, Rosiek,Slawianowska [23]; $\boxed{4.4,\ 4.5}$
enhanced FCNCs $d_i\, d_j\, \tilde{g}/\tilde{\chi}^0$	not accessible	$\boxed{4.4,\ 4.5,\ 5}$
vertex corrections $\bar{u}_{i,R}\, d_{j,L}\, H^+$	Degrassi,Gambino,Giudice [21]; Carena,Garcia, Nierste,Wagner [22];	process-dependent (non-universal)

Table 1.1: Summary of $\tan\beta$-enhanced effects with references to the literature. The boxes indicate new aspects which are for the first time discussed in this thesis with the numbers referring to the corresponding chapters.

missing in the derivation presented in [27]. Since the external leg method demands the inclusion of certain NLO diagrams into a LO calculation, it is not well suited for an implementation into computer programs like FeynArts [28] which are designed to perform fixed-order calculations. Therefore we develop an alternative resummation procedure in Section 4.5 based on matrix-valued wave function renormalisation. Using this approach, we derive effective Feynman rules in Section 4.5.2 which allow to account for $\tan\beta$-enhanced effects in LO calculations and can easily be implemented into computer programs. In Chapter 5 we finally investigate the consequences for rare B decays by studying the impact on the relevant effective Hamiltonian. We focus on gluino-squark contributions because they constitute novel corrections emerging from our treatment of $\tan\beta$-enhanced effects beyond the decoupling limit. We find a large effect in the Wilson coefficient C_{8g} of the chromomagnetic operator and demonstrate its importance by an analysis of the mixing-induced CP asymmetry in the decay $B \to \phi K_s$.

1.2. Probing new physics in electroweak penguins via hadronic B decays

Among the hadronic B decays, especially the $B \to K\pi$ modes are of great interest for investigating new physics. Not only are they dominated in the SM by FCNC loops and thus vulnerable to NP contributions. As a consequence of strong isospin symmetry, the dominating QCD penguin drops out if one considers ratios or differences of branching fractions and CP asymmetries of the four different decay channels $B^- \to \bar{K}^0\pi^-$, $B^- \to K^-\pi^0$, $\bar{B}^0 \to K^-\pi^+$ and $\bar{B}^0 \to \bar{K}^0\pi^0$. Therefore such observables are even sensitive to subleading isospin-violating effects. For example the ratios R_c and R_n of the branching fractions of the two charged B^- and the two neutral $\bar{B}^0$ decay modes are expected to satisfy $R_c \approx R_n$ within the SM [29]. The fact that experimental data showed for a long time a significant deviation from this pattern led to the formulation of a "$B \to K\pi$ puzzle" pointing to new physics in the electroweak penguin sector [30, 31]. As a consequence, $B \to K\pi$ decays have been studied in various elaborate works considering also CP violating observables and using different approaches to estimate non-perturbative effects. In Refs. [30–32] the hadronic $B \to K\pi$ parameters are related by $SU(3)$ flavour symmetry to their $B \to \pi\pi$ counterparts which are then taken from experimental data. Other analyses instead rely on a common fit to the $SU(3)$-related $B \to \pi\pi$ and $B \to K\pi$ data [33], some of them take into account $SU(3)$-breaking effects, too [34]. Moreover, as the experimental progress gave access to more and more $B \to K\pi$ observables, also a χ^2 fit to $B \to K\pi$ data alone became a popular strategy to determine the hadronic parameters [35]. Apart from parameter fits, also calculational tools like perturbative QCD [36], QCD factorisation [3, 37] or soft collinear effective theory [38] have been applied to $B \to K\pi$ decays. Almost all of the studies mentioned so far came to results in favour of new physics in the electroweak penguin sector.

In the meantime, however, the experimental values for the branching fractions have moved towards the SM predictions relaxing the CP conserving "$B \to K\pi$ puzzle" [39, 40]. On the other hand, the CP asymmetries still show some tensions which get manifest in the observable

$$\Delta A_{\mathrm{CP}} \equiv A_{\mathrm{CP}}(B^- \to K^-\pi^0) - A_{\mathrm{CP}}(\bar{B}^0 \to K^-\pi^+). \tag{1.1}$$

In the SM this quantity is expected to vanish to a very good approximation [41] whereas the current experimental value is given by [42]

$$\Delta A_{\mathrm{CP}} \overset{\mathrm{exp.}}{=} (14.8 \pm 2.8)\%. \tag{1.2}$$

Taking this discrepancy serious is very attractive as it may well be explained by new physics in electroweak penguins and many models predict large contributions in this sector [43, 44]. Examples for new physics scenarios which have been claimed to offer a solution to the "ΔA_{CP} puzzle" are models with a flavour-changing Z' coupling [45], fourth generation models [46], supersymmetry [47] and so on. On the other hand, ΔA_{CP} is sensitive to contributions from colour-suppressed tree-level diagrams as well. These contributions which come in QCDF with large hadronic uncertainties could as well lead to large deviations from a non-vanishing ΔA_{CP} and therefore the situation is not clear yet.

The scope of our work is as follows: Given the situation that tensions in $B \to K\pi$ decays nowadays essentially reside in one single observable which moreover suffers from large theoretical uncertainties, it is important to inspect also other hadronic decays which are sensitive to EW penguin contributions in order to decide whether there is really NP in this sector. Among these decays there are of course the $B \to K^*\pi, K\rho, K^*\rho$ channels which have the same flavour structure as the $B \to K\pi$ decays and differ only by the spin of the final state mesons. Even though experimental data on these decays is not as precisely determined as on their $B \to K\pi$ counterpart, they might give valuable constraints since they probe other chirality structures of the EW penguin operators than $B \to K\pi$ due to their different spin quantum numbers. Our main focus, however, is on the purely isospin-violating $B_s \to \phi\pi, \phi\rho$ decays. Being dominated by EW penguin contributions they constitute "golden modes" for probing this kind of NP. Apart from the suggestion by Fleischer [48] to use $B_s \to \phi\pi$ to measure the CKM angle γ, these decay modes have not drawn a lot of attention so far since they have not been observed yet due to their small branching fractions. However, an observation of $B_s \to \phi\rho$ at LHCb or $B_s \to \phi\pi$ at a potential Super-B factory is realistic, especially in presence of an enhancing NP contribution.

The plan of our analysis is as follows: In Chapter 6 we study the isospin structure of $B \to K\pi$ decays, construct observables which are sensitive to isospin violation and compare our SM results obtained with QCDF to the current experimental data. The discrepancy in ΔA_{CP} is then used as motivation to study the purely isospin-violating decays $B_s \to \phi\pi, \phi\rho$ whose structure and SM predictions are presented in Chapter 7. In Chapter 8 we analyse the consequences of NP in the EW penguin sector in a model-independent way: We add NP contributions to the Wilson coefficients $C_7^{(\prime)}, ..., C_{10}^{(\prime)}$, fit them to the $B \to K\pi$ data and study with respect to the resulting fit a potential enhancement of $B_s \to \phi\pi/\rho$. In Chapter 9 we perform an analogous study for two concrete types of NP models: Models with a modified, flavour-changing Z-boson coupling and models with a flavour-changing Z'-boson coupling. Finally, Chapter 10 summarises the most important results of this thesis.

2. General framework for the analysis of hadronic B decays

Rare hadronic B decays are an ideal place to search for new physics (NP) beyond the Standard Model (SM). This is because they are mediated at the quark level by flavour changing neutral currents (FCNCs) which are highly suppressed within the SM so that potential NP contributions have a chance to compete. Unfortunately, hadronic B decays are quite a challenge to theory: Due to the dominant low-energy QCD effects confining the quarks to hadrons, it is difficult to single out the high-energy FCNC transition which is responsible for the decay and which might contain NP effects.

In order to calculate transition amplitudes for hadronic B decays it is essential to make use of the strong hierarchy $v \gg m_b \gg \Lambda_{\rm QCD}$ between the three energy scales involved: the scale of the FCNC transition and potential NP, established by the electroweak vacuum expectation value (vev) $v \sim 175$ GeV, the scale of the b-quark mass $m_b \sim 4.2\,\rm GeV$ and finally the fundamental scale of QCD $\Lambda_{\rm QCD} \sim 350$ MeV governing the size of non-perturbative effects. In a first step the hierarchy $v \gg m_b$ is exploited to construct a simplified effective theory valid around the scale m_b. To determine the hadronic matrix elements of the corresponding effective Hamiltonian, one can then in a second step take advantage of the hierarchy $m_b \gg \Lambda_{\rm QCD}$. It allows to separate the perturbative part of the matrix elements from the non-perturbative one via the QCD factorisation approach (QCDF) developed by Beneke, Buchalla, Neubert and Sachrajda (BBNS approach) [3]. Non-perturbative interactions are then confined to decay constants and form factors, i.e. to a minimal number of quantities.

This chapter introduces the reader to the basic concepts of these methods. At the same time it serves to define the conventions, schemes and notation used throughout the thesis. After setting up the effective Hamiltonian for $\Delta B = \Delta S = 1$ hadronic B decays in section 2.1, the calculation of its matrix elements via the QCDF approach is discussed in section 2.2.

2.1. The effective $\Delta B = \Delta S = 1$ Hamiltonian

2.1.1. Definition and construction

Rare hadronic B decays are induced by a flavour-changing weak transition of the constituent b quark of the B meson. The released energy $E = m_B$ is much too low for the production of the top quark, the W-, Z- and Higgs-bosons and potential NP particles. For this reason it suffices to consider a simplified theory which only contains the five light quarks, the gluon and the photon as dynamical fields and reproduces the S-matrix elements of the full theory (SM plus NP) in the low-energy region. The corresponding effective Hamiltonian contains in addition to the usual QED and QCD terms (reduced to five active flavours) also higher-dimensional operators encoding the flavour changing weak transitions. In the case of $\Delta B = \Delta S = 1$ decays and with the SM as underlying full theory this higher dimensional part of the Hamiltonian reads

$$\mathcal{H}_{\text{eff}}^{(1)} = \frac{4G_F}{\sqrt{2}} \sum_{p=u,c} \lambda_p^{(s)} \left(C_1 Q_1^p + C_2 Q_2^p + \sum_{i=3}^{10} C_i Q_i + C_{7\gamma} Q_{7\gamma} + C_{8g} Q_{8g} \right) + \text{h.c.}\,, \tag{2.1}$$

where G_F is Fermi's constant and $\lambda_p^{(s)} = V_{pb} V_{ps}^*$ represents a product of elements of the quark mixing (CKM) matrix. Explicit expressions for the current-current operators $Q_{1,2}^p$, QCD penguin operators $Q_{3,\ldots,6}$, electroweak penguin operators $Q_{7,\ldots,10}$ and the electromagnetic and chromo-magnetic operators $Q_{7\gamma}$, Q_{8g} are given by

$$\begin{aligned}
Q_1^p &= (\bar{s}_\alpha \gamma^\mu P_L p_\alpha)(\bar{p}_\beta \gamma_\mu P_L b_\beta)\,, & Q_2^p &= (\bar{s}_\alpha \gamma^\mu P_L p_\beta)(\bar{p}_\beta \gamma_\mu P_L b_\alpha)\,, \\
Q_3 &= (\bar{s}_\alpha \gamma^\mu P_L b_\alpha) \sum_q (\bar{q}_\beta \gamma_\mu P_L q_\beta)\,, & Q_4 &= (\bar{s}_\alpha \gamma^\mu P_L b_\beta) \sum_q (\bar{q}_\beta \gamma_\mu P_L q_\alpha)\,, \\
Q_5 &= (\bar{s}_\alpha \gamma^\mu P_L b_\alpha) \sum_q (\bar{q}_\beta \gamma_\mu P_R q_\beta)\,, & Q_6 &= (\bar{s}_\alpha \gamma^\mu P_L b_\beta) \sum_q (\bar{q}_\beta \gamma_\mu P_R q_\alpha)\,, \\
Q_7 &= (\bar{s}_\alpha \gamma^\mu P_L b_\alpha) \sum_q \frac{3}{2} e_q (\bar{q}_\beta \gamma_\mu P_R q_\beta)\,, & Q_8 &= (\bar{s}_\alpha \gamma^\mu P_L b_\beta) \sum_q \frac{3}{2} e_q (\bar{q}_\beta \gamma_\mu P_R q_\alpha)\,, \\
Q_9 &= (\bar{s}_\alpha \gamma^\mu P_L b_\alpha) \sum_q \frac{3}{2} e_q (\bar{q}_\beta \gamma_\mu P_L q_\beta)\,, & Q_{10} &= (\bar{s}_\alpha \gamma^\mu P_L b_\beta) \sum_q \frac{3}{2} e_q (\bar{q}_\beta \gamma_\mu P_L q_\alpha)\,, \\
Q_{7\gamma} &= \frac{e}{16\pi^2} m_b (\bar{s}_\alpha \sigma^{\mu\nu} P_R b_\alpha) F_{\mu\nu}\,, & Q_{8g} &= \frac{g_s}{16\pi^2} m_b (\bar{s}_\alpha \sigma^{\mu\nu} T^a_{\alpha\beta} P_R b_\beta) G^a_{\mu\nu}\,.
\end{aligned} \tag{2.2}$$

with e_q denoting the quark charges in units of $|e|$, α, β being colour indices and the sum extending over $q = u, d, s, c, b$.

The Hamiltonian (2.1) describes point-like interactions among the photon, gluons and light quarks, with the Wilson coefficients C_i being the corresponding coupling constants. Energies accessible in B decays are too low to resolve the inner structure of these point-like vertices, i.e. the exchanged heavy virtual particles like the W-boson. From the point of view of the low-energy theory, the couplings C_i are just free parameters to be determined from experiment. However, knowing the underlying full theory, one can relate the C_i to the more fundamental parameters of the full theory by performing a so-called matching calculation, i.e. by calculating S-matrix elements in both theories and requiring equality of the results.

In writing down the Hamiltonian in (2.1) the SM has implicitly been assumed to be the full theory behind it. In general one would have to write down all operators obeying the $SU(3)_{\mathrm{C}}$ and $U(1)_{\mathrm{em}}$ symmetries of QCD and QED. The restriction to the ones in (2.2) results from the specific structure of the Yukawa matrices which are the only sources of flavour violation in the SM: The hierarchical texture of the diagonalised Yukawa matrices imply that operators generated only at the expense of a Yukawa coupling other then the one of the top quark can be dropped from the Hamiltonian. In the first part of this thesis, we study a specific scenario of the Minimal Supersymmetric Standard Model (MSSM) called Minimal Flavour Violation (MFV). Since this scenario is defined by requiring that the Yukawa matrices are the only source of flavour violation, it exhibits the same flavour structure as the SM and so its effective Hamiltonian is spanned by the same operator basis (2.2)[1]. In the second part of the thesis, a model-independent analysis is performed. Also in this case we assume that the flavour structure of NP can still be described in terms of the SM operators, however possibly with addition of the so-called "mirror" operators Q_i', obtained from the Q_i by a global exchange of left- and right chiralities of the quark fields. This hypothesis is fulfilled within many realistic NP models.

2.1.2. Renormalisation group evolution

Since the effective theory has been constructed to reproduce the low-energy limit of the full theory, both have the same infrared behaviour. However, due to the absence of heavy particles and their propagators in the effective theory, the ultraviolet structure is different. In the effective theory additional divergences emerge which have to be absorbed by renormalising the effective couplings C_i. Exactly as in the case of the strong coupling α_s, the renormalisation leads to a scale dependence of the renormalised couplings which can be cast into a coupled set of differential equations

$$\mu \frac{d}{d\mu} C_i(\mu) = \gamma_{ij}^T C_j(\mu) , \tag{2.3}$$

the so-called renormalisation group equations (RGEs). Including next-to-leading order (NLO) contributions of the form α_s^2 and $\alpha_s \, \alpha_e$ with α_e denoting the fine-structure constant, the anoma-

[1] Actually, in the MSSM with MFV there are new scalar operators involving a pair of b quarks. These operators originate from a possible enhancement of the down-type Yukawa couplings such that the b quark Yukawa coupling cannot be neglected anymore. However, as studied in Ref. [49], their effects are tightly constrained from $B_s \to \mu^+\mu^-$ and $B \to \tau\nu$.

	C_1	C_2	C_3	C_4	C_5	C_6
LO	1.118	−0.270	0.012	−0.028	0.008	−0.034
NLO	1.082	−0.192	0.014	−0.036	0.009	−0.042
	C_7/α_e	C_8/α_e	C_9/α_e	C_{10}/α_e	$C_{7\gamma}^{\rm eff}$	$C_{8g}^{\rm eff}$
LO	−0.005	0.028	−1.289	0.292	−0.317	−0.151
NLO	−0.016	0.059	−1.268	0.229	—	—

Table 2.1: Short-distance coefficients at the scale $\mu = m_b$. The RGE evolution follows the scheme described in the text using $\Lambda_{\rm QCD}^{(5)} = 0.231\,{\rm GeV}$, $m_t(m_t) = 161.45\,{\rm GeV}$, $m_b(m_b) = 4.2\,{\rm GeV}$, $M_W = 80.4\,{\rm GeV}$, $M_Z = 91.2\,{\rm GeV}$ and $\alpha_e = 1/129$.

lous dimension matrix γ is of the form

$$\gamma = \frac{\alpha_s}{4\pi}\gamma_s^{(0)} + \frac{\alpha_e}{4\pi}\gamma_e^{(0)} + \left(\frac{\alpha_s}{4\pi}\right)^2\gamma_s^{(1)} + \frac{\alpha_s}{4\pi}\frac{\alpha_e}{4\pi}\gamma_{se}^{(1)}. \tag{2.4}$$

For explicit expressions of $\gamma^{(0)}$, $\gamma_e^{(0)}$, $\gamma_s^{(1)}$ and $\gamma_{se}^{(1)}$, we refer to Ref. [50] in case of the four-quark operators. Since $Q_{7\gamma}$ and Q_{8g} enter four-quark amplitudes only at NLO, a leading order (LO) treatment of $C_{7\gamma}$ and C_{8g} is sufficient for the study of hadronic B decays. The relevant entries of $\gamma_s^{(0)}$ involving $C_{7\gamma}$ and C_{8g} can be found in Ref. [51]. With respect to the anomalous dimension (2.4), the solution of the RGEs (2.3) reads

$$\begin{aligned} &C_i(\mu) = U_{ij}^{(1)}(\mu,\mu_0)\,C_j(\mu_0), \qquad U^{(1)}(\mu,\mu_0) = W(\mu)\,U^{(0)}(\mu,\mu_0)\,W^{-1}(\mu_0),\\ &\text{with} \quad W(\mu) = \left(1+\frac{\alpha_e}{4\pi}K\right)\left(1+\frac{\alpha_s(\mu)}{4\pi}J\right)\left(1+\frac{\alpha_e}{\alpha_s(\mu)}P\right),\\ &\text{and} \quad U^{(0)}(\mu,\mu_0) = \left(\frac{\alpha_s(\mu_0)}{\alpha_s(\mu)}\right)^{\frac{\gamma_s^{(0)T}}{2\beta_0}}. \end{aligned} \tag{2.5}$$

Explicit expressions for J, P and K can easily be determined by solving algebraic equations, obtained from inserting (2.5) into (2.3) [50]. Given the Wilson coefficients $C_i(\mu_0)$ at a certain energy scale μ_0, equation (2.5) can then be used to calculate their value at any other scale μ. Since QCD and QED are invariant under the parity transformation, the coefficients C_i' of the mirror operators obey the same RGEs as their unprimed counterparts.

The accuracy of the effective theory is given by the precision with which the Wilson coefficients C_i are determined in the matching procedure. The result of a perturbative calculation of the C_i at the scale μ involves logarithms of the form $\alpha_s\,\ln(\mu/M_W)$. At the scale $\mu \sim m_b$ at which the effective Hamiltonian is applied to the hadronic B decay in question, these logarithms become large spoiling the perturbative expansion. However, the RGE technique can be used to bypass this problem: the trick is to calculate the Wilson coefficients C_i at the scale $\mu_0 \sim M_W$ at which the dangerous logarithms tend to vanish and then to use the RGE evolution (2.5) to evolve them down to the scale of interest $\mu \sim m_b$. In this way the large logarithms $(\alpha_s \ln(\mu/M_W))^n$ get resummed to all orders $n = 1, 2, \ldots$.

Now, after the general method has been described, we want to specify the modified scheme on which we rely for our analyses. As we will discuss in the next section, matrix elements of the effective Hamiltonian are evaluated using the QCDF method developed in Ref. [3]. This evaluation amounts to a NLO calculation with some simplifications, e.g. $\mathcal{O}(\alpha_e)$ corrections to the matrix elements of the operators $Q_1, ..., Q_6$ are neglected. For consistency, the same simplifications have to be applied to the NLO initial conditions and the RGE evolution for the Wilson coefficients adapting the scheme defined in Ref. [3]: In the initial conditions for $C_1, ..., C_6$, corrections of $\mathcal{O}(\alpha_e)$ are neglected, $\mathcal{O}(\alpha_s)$ mixing of $C_7, ..., C_{10}$ into $C_1, ..., C_6$ is switched off in $\gamma_s^{(1)}$, and $\mathcal{O}(\alpha_e)$ mixing in $\gamma_{se}^{(1)}$ is taken into account only for the mixing of $C_1, ..., C_6$ into $C_7, ..., C_{10}$.

In the second part of the thesis, it is our goal to investigate effects of new physics in the electroweak penguin sector entering the coefficients $C_7, \ldots, C_{10}$ and $C'_7, \ldots, C'_{10}$. In the SM, contributions to $C_7, \ldots, C_{10}$ arise only at $\mathcal{O}(\alpha_e)$, albeit partly enhanced by factors $x_{tW} = m_t^2/M_W^2$ and/or $1/\sin^2\theta_W$. Following the approach of Ref. [3] we treat the enhanced parts as LO in the RGE evolution. To be consistent we neglect at the same time any mixing of $C_7, \ldots, C_{10}$ into $C_1, ..., C_6$. Since this treatment improves the RGE evolution for $C_7, ..., C_{10}$ and since it is exactly these coefficients we are interested in, it is well suited for our analysis. The values for the SM coefficients obtained in this scheme are given in table 2.1. By contrast, for the NP contributions we use the standard treatment for the LO RGE.

2.2. QCD factorisation

In the last section, we have constructed an effective Hamiltonian $\mathcal{H}_{\text{eff}}^{(1)}$ valid at the scale $\mu \sim m_b$. This effective Hamiltonian can now be used to calculate amplitudes for hadronic B decays. The amplitude for the two-body decay $B \to M_1 M_2$ is then given by the corresponding matrix element of $\mathcal{H}_{\text{eff}}^{(1)}$. Schematically written, it reads

$$\mathcal{A}(B \to M_1 M_2) = \frac{4G_F}{\sqrt{2}} \sum_i \lambda_i \, C_i(\mu) \, \langle M_1 M_2 | Q_i | B \rangle(\mu) \tag{2.6}$$

where Q_i denote the effective operators contained in $\mathcal{H}_{\text{eff}}^{(1)}$ with C_i and λ_i being the associated Wilson coefficients and CKM factors. The task is now to determine the matrix elements

$\langle M_1 M_2 | Q_i | B \rangle(\mu)$ which encode the information on how the quarks building the operator Q_i hadronise into the mesons B, M_1 and M_2. Therefore, to calculate these matrix elements one has to rely on non-perturbative methods, such as lattice gauge theory or QCD sum rules. However, before this is done, one can, in the case where the mesons M_1, M_2 are light compared to the B-meson, employ the special kinematic of the decay to separate parts of the matrix elements which can still be treated perturbatively from the non-perturbative parts which thereby factorise into simpler objects, namely form factors and decay constants[2].

2.2.1. The factorisation formula

Physically the factorisation property of B decays into light mesons mentioned above can be understood as a consequence of colour transparency [52]. The b-quark decays into three light quarks, one of them forming together with the spectator quark the meson M_1, the other two hadronising into the meson M_2. These light quarks are very energetic because $m_b \gg \Lambda_{\text{QCD}}$ and they originate from a common space-time point since they are generated via a point-like interaction. Thus to hadronise into M_2 the corresponding two quarks have to be highly collinear constituting a colour neutral system of small transverse extension. Therefore, this two-quark system is "invisible" (colour-transparent) to a soft gluon whose energy is too low to resolve its inner structure. As a consequence, non-perturbative strong interactions are confined to the B-M_1 system and to the M_2 system separately, so that the corresponding parts of the matrix element factorise (left pictogram of fig. 2.1). Interactions between these two sub-systems are only due to hard gluon exchange and can be calculated by usual perturbation theory. On the other hand, the quark system forming the meson M_1 will not generally factorise from the one forming the original B meson since the spectator quark ending up in M_1 is low-energetic. An exception occurs in the case when the spectator undergoes a hard gluon scattering before hadronising. This situation in which the matrix element factorises into three form factors is displayed in the right pictogram of fig. 2.1.

The diagrams of fig. 2.1 can be cast into a factorisation formula [3]

$$\begin{aligned}\langle M_1 M_2 | Q_i | \bar{B} \rangle &= \sum_j F_j^{B\to M_1}(m_2^2) \int_0^1 du_2\, T_{ij}^I(u_2)\, \Phi_{M_2}(u_2) \;+\; (M_1 \leftrightarrow M_2) \\ &\quad + \int_0^1 du_B\, du_1\, du_2\, T_i^{II}(u_B, u_1, u_2)\, \Phi_B(u_B)\, \Phi_{M_1}(u_1)\, \Phi_{M_2}(u_2), \qquad (2.7)\end{aligned}$$

valid up to corrections of order $\mathcal{O}(\Lambda_{\text{QCD}}/m_b)$. Here, $u_B, u_1, u_2 \in (0,1)$ stand for the longitudinal momentum fractions carried by the valence quarks in the mesons B, M_1, M_2, i.e. labelling the

[2] It should be mentioned that factorisation also holds in the case where the meson which picks up the spectator quark is heavy. However, we will not consider this option here since we are only interested in B decays into two light mesons.

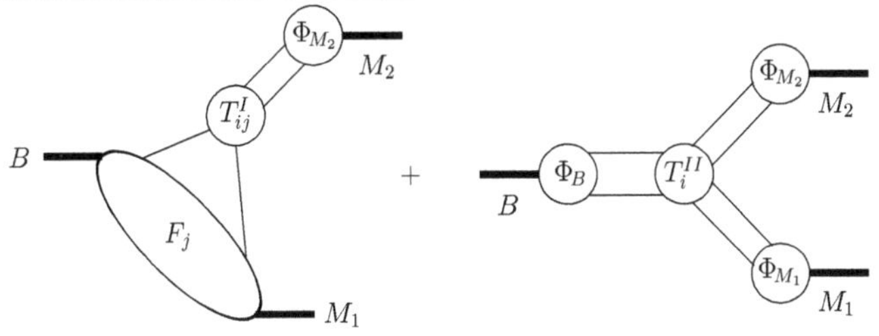

Figure 2.1: Graphical representation of the factorisation formula. Only one of the two form-factor terms in (2.7) is shown for simplicity. Picture taken from [3].

momenta of the mesons by p_i the longitudinal momenta of the valence quarks are given by $u_i p_i$ whereas those of the anti-quarks are given by $(1-u_i)p_i$. The hard scattering kernels

$$T^I_{ij}(u_2) = 1 + \mathcal{O}(\alpha_s(\mu_b)) \quad \text{and} \quad T^{II}_i(u_B, u_1, u_2) = \mathcal{O}(\alpha_s(\mu_h)) \tag{2.8}$$

with $\mu_b \sim m_b$ and $\mu_h \sim \sqrt{\Lambda_{\rm QCD} m_b}$ are calculated perturbatively. They are weighted with the probabilities $\Phi_{M_i}(u_i)$ for the quarks to carry the corresponding momentum fractions u_i inside the mesons before an integration over all possible momentum configurations is performed. The probabilities $\Phi_{M_i}(u_i)$ are given by the so-called light-cone distribution amplitudes (LCDAs) which are in complete analogy to the parton distribution functions in collider physics. In the heavy quark limit $\Lambda_{\rm QCD} \ll m_b$ it suffices to consider the LCDAs within the leading twist approximation, i.e. higher Fock states involving sea-quarks and -gluons can be neglected. Finally, $F_j^{B\to M}(q^2)$ are the usual form factors parametrising the matrix element $\langle M_1|\bar q\,\Gamma_j\, b|\bar B_q\rangle$ to be evaluated at $p^2 = m_2^2 \approx 0$ where Γ_j parametrises the particular Dirac structure and m_2 is the mass of the meson M_2.

From (2.8) we infer that at $\mathcal{O}(\alpha_s^0)$ the second line in Eq. (2.7) vanishes and the integral in the first line reduces to the decay constant f_{M_2} of the light meson M_2 which is the proper normalisation constant for the LCDA. So in this case we recover the naive factorisation result for the amplitude (2.6) with the matrix elements evaluated as

$$\begin{aligned} \langle M_1 M_2|\, Q_i\,|\bar B_q\rangle^{(\rm F)} &= \langle M_1|\,\bar q\,\Gamma\, b\,|\bar B_q\rangle\,\langle M_2|\,\bar q\,\Gamma\, q'\,|0\rangle + (M_1 \leftrightarrow M_2) \\ &\equiv \langle Q_i\rangle^{(\rm F)}_{12} + \langle Q_i\rangle^{(\rm F)}_{21}\,. \end{aligned} \tag{2.9}$$

Going beyond LO in α_s, the factorisation formula guarantees that the amplitude (2.6) can still be

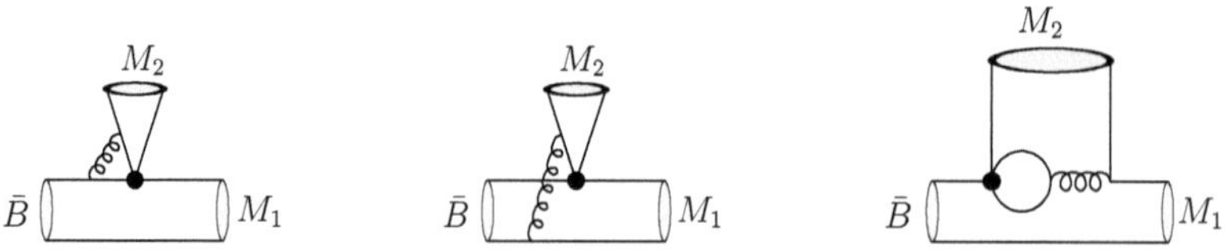

Figure 2.2: Perturbative non-factorisable QCD corrections contributing (from left to right) to V_i, H_i and P_i^p

expressed in terms of the factorised matrix elements, i.e. it is of the form

$$\mathcal{A}(\bar{B} \to M_1 M_2) = \frac{4G_F}{\sqrt{2}} \sum_{p=u,c} \lambda_p^{(s)} \sum_{i=1}^{10} \left(a_i^p(M_1M_2) \langle Q_i \rangle_{12}^{(\mathrm{F})} + a_i^p(M_2M_1) \langle Q_i \rangle_{21}^{(\mathrm{F})} \right). \quad (2.10)$$

However, the coefficients are now not simply given by the Wilson coefficients C_i anymore but they include perturbative power corrections of order $\alpha_s(\mu_{b,h})$:

$$\begin{aligned} a_i^p(M_1M_2) &= \left(C_i + \frac{C_{i\pm1}}{N_c} \right) N_i(M_2) + \\ &\quad C_{i\pm1} \frac{C_F}{N_c} \left[\frac{\alpha_s(\mu_b)}{4\pi} V_i(M_2) + \frac{4\pi^2}{N_c} \frac{\alpha_s(\mu_h)}{4\pi} H_i(M_1M_2) \right] + P_i^p(M_2), \\ a_1^c(M_1M_2) &= a_2^c(M_1M_2) = 0, \end{aligned} \quad (2.11)$$

where the upper (lower) sign applies when i is odd (even). The first line in (2.11) is the tree-level contribution leading to the naive factorisation result. The quantities V_i, H_i and P_i^p in the second line represent the vertex corrections, the hard spectator interactions and the penguin contractions, respectively. Each type of correction is illustrated in Fig. 2.2 by a corresponding Feynman diagram. Explicit formulae for V_i, H_i and P_i^p depend on the final state, whether it is a PP, PV, VP or VV combination with P denoting pseudoscalar and V denoting vector mesons, and can be found in Refs. [3, 5]. Note that the operators $Q_{7\gamma}$ and Q_{8g} do not appear in (2.10) because their tree level matrix elements vanish for hadronic B decays. However, they contribute to the penguin coefficients P_i^p. It should also be remarked that unlike the Wilson coefficients in effective theories the coefficients $a_i^p(M_1M_2)$ are not process-independent but depend on the hadrons in the final state via the LCDAs Φ_{M_i}.

It is useful to combine amplitudes a_i whose operators Q_i involve the same flavour and colour but a different chirality structure to flavour amplitudes α_k. The chirality structure is responsible for the fact that the pattern according to which the operators Q_i enter the decay amplitudes depends on the spin of the final-state mesons and is thus different for PP, PV, VP and VV decays. Hiding the chirality structure in the flavour amplitudes α_k therefore allows a universal description of all these types of decays in terms of the α_k. We give here explicitly the expressions for the flavour amplitudes related to the electroweak penguin operators, namely

$$\alpha^p_{3\rm EW}(M_1M_2) = \begin{cases} a^p_9(M_1M_2) - a^p_7(M_1M_2), & \text{if } M_1M_2 = PP, VP \\ a^p_9(M_1M_2) + a^p_7(M_1M_2), & \text{if } M_1M_2 = PV, VV \end{cases},$$

$$\alpha^p_{4\rm EW}(M_1M_2) = \begin{cases} a^p_{10}(M_1M_2) + r^{M_2}_\chi\, a^p_8(M_1M_2), & \text{if } M_1M_2 = PP, PV \\ a^p_{10}(M_1M_2) - r^{M_2}_\chi\, a^p_8(M_1M_2), & \text{if } M_1M_2 = VP, VV \end{cases} \tag{2.12}$$

since they will play a major role in part II of this work. The corresponding QCD penguin amplitudes α^p_3 and α^p_4 are obtained from $\alpha^p_{3,\rm EW}$ and $\alpha^p_{4,\rm EW}$ by the replacements $a^p_9 \to a^p_3$, $a^p_7 \to a^p_5$, $a^p_{10} \to a^p_4$ and $a^p_8 \to a^p_6$ whereas the current-current amplitudes are simply given by $\alpha^u_1 = a^u_1$ and $\alpha^u_2 = a^u_2$. The a^p_6- and a^p_8-terms vanish at leading twist in the LCDAs since the matrix elements $\langle M_1M_2|\, Q_{6,8}\, |\bar B_q\rangle^{(\rm F)}$ multiplying them involve (pseudo-)scalar vacuum-meson currents[3]. The twist-3 contribution is formally $\Lambda_{\rm QCD}/m_b$-suppressed. However, due to a chiral enhancement of the normalisation ratio $r^{M_2}_\chi$, e.g. $m_\pi/m_{u,d}$ for $M_2 = \pi$, it can gain numerical relevance. Explicit expressions for the $r^{M_2}_\chi$ can be found in [3,5].

As mentioned in Section 2.1.1 we want to study also contributions from the mirror operators Q'_i. To this end we have to extend Eq. (2.10) by corresponding mirror-terms involving further amplitudes a'^p_i. Since QCD is parity-invariant, the NLO expression for the a'^p_i is equivalent to the one for a^p_i in Eq. (2.11), only the short distance coefficients C_k have to be replaced by the C'_k. Applying further a parity transformation to the matrix elements

$$\langle M_1M_2|Q_i|\bar B\rangle = -\eta_{M_1M_2}\, \langle M_1M_2|Q'_i|\bar B\rangle \tag{2.13}$$

one concludes that [4]

$$\mathcal{A}^p_i(M_1M_2) \propto a^p_i(M_1, M_2) - \eta_{M_1M_2}\, a'^p_i(M_1, M_2), \tag{2.14}$$

where $\eta_{M_1M_2} = \pm 1$ is the parity of the final state. This observation can be used to translate Eq. (2.12) to the mirror sector. With $\eta_{M_1M_2} = 1$ for PP, VV and $\eta_{M_1M_2} = -1$ for PV we

[3]To see this one has to perform a Fierz transformation on $Q_{6,8}$ bringing them into the form of two colour-neutral bilinears.

obtain [49]

$$\alpha'^{p}_{3\mathrm{EW}}(M_1M_2) = \begin{cases} -a'^{p}_9(M_1M_2) + a'^{p}_7(M_1M_2), & \text{if } M_1M_2 = PP \\ a'^{p}_9(M_1M_2) + a'^{p}_7(M_1M_2), & \text{if } M_1M_2 = PV \\ a'^{p}_9(M_1M_2) - a'^{p}_7(M_1M_2), & \text{if } M_1M_2 = VP \\ -a'^{p}_9(M_1M_2) - a'^{p}_7(M_1M_2), & \text{if } M_1M_2 = VV \end{cases},$$

$$\alpha'^{p}_{4\mathrm{EW}}(M_1M_2) = \begin{cases} -a'^{p}_{10}(M_1M_2) - r_\chi^{M_2}\, a'^{p}_8(M_1M_2), & \text{if } M_1M_2 = PP \\ a'^{p}_{10}(M_1M_2) + r_\chi^{M_2}\, a'^{p}_8(M_1M_2), & \text{if } M_1M_2 = PV \\ a'^{p}_{10}(M_1M_2) - r_\chi^{M_2}\, a'^{p}_8(M_1M_2), & \text{if } M_1M_2 = VP \\ -a'^{p}_{10}(M_1M_2) + r_\chi^{M_2}\, a'^{p}_8(M_1M_2), & \text{if } M_1M_2 = VV \end{cases}. \tag{2.15}$$

Furthermore, in order to find the appropriate formula for the primed amplitude of a $B \to VV$ decay, any quantity in the expression for the unprimed amplitude which depends on the helicity of a vector meson has to be replaced by the one involving the opposite helicity.

Finally, we should briefly comment on contributions to the decay amplitudes arising from diagrams in which the two valence quarks of the B meson annihilate. These weak annihilation contributions are subleading in $\Lambda_{\mathrm{QCD}}/m_b$ and therefore they do not appear in the factorisation formula (2.7). In the BBNS approach of QCDF their effects are estimated and promoted to flavour amplitudes β_i which add to the α_i. For a detailed discussion and explicit expressions of the β_i we refer to Refs. [3,5].

2.2.2. Properties and limitations of QCDF

Having discussed the basic concept of the QCDF framework, we summarise in this section important properties and limitations of this approach which one should keep in mind when applying it to phenomenological analyses:

- The overall concept of QCDF is valid in the heavy-quark limit $m_b \to \infty$. Thus it is expected to give correct results only up to order $\Lambda_{\mathrm{QCD}}/m_b$ corrections. This expansion parameter is numerically not much smaller than α_s. Therefore predictions for decays where the LO amplitude is small and which are therefore very sensitive to $\mathcal{O}(\alpha_s)$-effects suffer from large uncertainties.

- One class of such processes are decays with a colour-suppressed LO topology. This means that the colour-indices α and β of the two quark chains of the dominantly contributing operator have to be equal in order to form colour-neutral final-state mesons. This leads to a $1/N_c$ suppression of the LO amplitude with $N_c = 3$ being the number of colours which can be lifted by $\mathcal{O}(\alpha_s)$ corrections. As discussed in the first point, this implies large uncertainties for QCDF predictions of such decays.

- Since form factors and decay constants are real, the operator matrix elements are real at LO in QCDF. As a consequence, strong phases are in QCDF either perturbative and suppressed by α_s or non-perturbative and suppressed by $\Lambda_{\rm QCD}/m_b$. Hence QCDF predicts rather small strong phases which are subject to large uncertainties from $\Lambda_{\rm QCD}/m_b$ corrections. The largest phases are expected in amplitudes with a suppressed LO contribution.

- In $B \to VV$ decays the final-state vector mesons can either be both longitudinally or both positively or both negatively polarised. Because of the left-handed nature of weak interactions in the SM, helicity-flips of the fast travelling light quarks are needed in order to form transversely polarised vector mesons, in fact one flip in the case of negative and two flips in the case of positive helicities [4]. This causes a hierarchy

$$\mathcal{A}_0 : \mathcal{A}_- : \mathcal{A}_+ = 1 : \frac{\Lambda_{\rm QCD}}{m_b} : \left(\frac{\Lambda_{\rm QCD}}{m_b}\right)^2 \tag{2.16}$$

of the corresponding amplitudes at LO in QCDF. Being already subleading in $\Lambda_{\rm QCD}/m_b$ the calculation of negative helicity amplitudes raises conceptual problems resulting in large uncertainties of the total branching fractions. This can be avoided by considering solely decays into longitudinal vector mesons which, however, require the measurement of angular distributions. Positively polarised amplitudes can be neglected in QCDF to a good precision. Further details are discussed in Ref. [5]. Here we remark only that mirror contributions develop of course the reversed hierarchy

$$\mathcal{A}'_0 : \mathcal{A}'_- : \mathcal{A}'_+ = 1 : \left(\frac{\Lambda_{\rm QCD}}{m_b}\right)^2 : \frac{\Lambda_{\rm QCD}}{m_b} \tag{2.17}$$

suggesting polarisation measurements as a possible tool for probing right-handed new physics [4].

2.2.3. Input parameters

In table 2.2 we display the numerical values which we use for the theoretical input parameters throughout this work. The values for the QCD and CKM parameters as well as the B meson lifetimes are taken from [53]. The value m_q of the light quark masses is determined in such a way that $r_\chi^K = r_\chi^\pi$ for the ratios introduced in Eq. (2.12). The hadronic parameters consist of the decay constants and form factors into which the factorised operator matrix elements decompose and the Gegenbauer moments which parametrise the LCDAs. Here we use the most updated values from Refs. [3, 5, 54]. For our numerical calculations we add errors in quadrature and vary the scale μ_b at which the decay amplitudes are calculated between $m_b/2$ and $2m_b$.

In the BBNS approach of QCDF some $\Lambda_{\rm QCD}/m_b$ suppressed but numerically sizable contributions in the hard scattering amplitudes as well as the weak annihilation amplitudes are estimated by introducing unknown $\mathcal{O}(1)$ parameters $X_{H,A,L}$. For phenomenological analyses they are parametrised as

$$X_{H,A} = (1 + \rho_{H,A} e^{i\varphi_{H,A}}) \ln \frac{m_B}{\Lambda_h}, \qquad X_L = (1 + \rho_L e^{i\varphi_L}) \frac{m_B}{\Lambda_h}, \tag{2.18}$$

where the ρ_i are real, φ_i are arbitrary strong phases, m_B denotes the B meson mass and $\Lambda_h = 0.5\,\mathrm{GeV}$. We will assume $\rho_{H,L} = 0$ by default and consider $\rho_{H,L} \leq 1$ and arbitrary phases $\varphi_{H,L}$ in the error estimation. Concerning X_A we take

$$\rho_A = \begin{cases} 1.5, & \text{if } M_1M_2 = PP, PV, VP, \\ 1, & \text{if } M_1M_2 = VV \end{cases}, \qquad \varphi_A = \begin{cases} -55^\circ, & \text{if } M_1M_2 = PP, \\ -20^\circ, & \text{if } M_1M_2 = PV, \\ -70^\circ, & \text{if } M_1M_2 = VP, \\ -40^\circ, & \text{if } M_1M_2 = VV \end{cases} \tag{2.19}$$

as default values. For the error determination we vary $\rho_A \leq 1.5$ in the PP, PV, VP case, $\rho_A \leq 1$ in the VV case and $-180^\circ < \varphi_A \leq 180^\circ$ in all cases. This corresponds to "Scenario S4" defined in the last paper of [3] apart from one exception: Given the current data, the value for ρ_A in the standard "Scenario S4", namely $\rho_A = 1$ for PP, PV, VP and $\rho_A = 0.6$ for VV, generates a too small result for the ratio of the $B^- \to \bar{K}^0\pi^-$ and $B^- \to \pi^-\pi^0$ branching fractions. These decays are dominated by QCD penguin and tree contributions, respectively, and are expected to receive negligible contributions from NP. In order to have a conservative estimate of the theory error, we have therefore increased the default value and the error interval for the parameter ρ_A.

QCD scale and running quark masses [GeV]				
$\Lambda_{\overline{\rm MS}}^{(5)}$	$m_b(m_b)$	$m_c(m_b)$	$m_s(2\,\mathrm{GeV})$	m_q/m_s
0.231	4.2	1.3±0.2	0.090± 0.020	0.0413
CKM parameters				
λ	$\lvert V_{cb}\rvert$	$\lvert V_{ub}/V_{cb}\rvert$	γ	$\sin(2\beta)$
0.225	0.0415 ± 0.0010	$0.085^{+0.025}_{-0.015}$	$(70 \pm 10)^\circ$	0.673 ± 0.23
B meson parameters				
		B^-	$\bar{B}^0$	$\bar{B}^0_s$
Lifetime	τ[ps]	1.638	1.525	1.472
Decay constant	f_B[MeV]	210 ± 20		240 ± 20
	λ_B[MeV]	200^{+250}_{-0}		200^{+250}_{-0}

Pseudoscalar-meson decay constants and Gegenbauer moments		
	π	K
f[MeV]	131	160
$a_1, a_{1,\perp}$	0	0.06 ± 0.06
$a_2, a_{2,\perp}$	0.20 ± 0.15	0.20 ± 0.15

Vector-meson decay constants and Gegenbauer moments			
	ρ	K^*	ϕ
f[MeV]	209 ± 1	218 ± 4	221 ± 3
$f^\perp$[MeV]	165 ± 9	185 ± 10	186 ± 9
$a_1, a_{1,\perp}$	0	0.06 ± 0.06	0
$a_2, a_{2,\perp}$[MeV]	0.1 ± 0.2	0.1 ± 0.2	0 ± 0.3
Pseudoscalar-meson form factor at $q^2 = 0$			
	$B \to \pi$	$B \to K$	$B_s \to \bar{K}$
f_+	0.25 ± 0.05	0.34 ± 0.05	0.31 ± 0.05
Vector-meson form factor at $q^2 = 0$			
	$B \to \rho$	$B \to K^*$	$B_s \to \phi$
A_0	$0.30^{+0.07}_{-0.03}$	0.39 ± 0.06	$0.38^{+0.10}_{-0.02}$
f_+	0.00 ± 0.06	0.00 ± 0.06	0.00 ± 0.06
f_-	0.55 ± 0.06	0.68 ± 0.07	$0.65^{+0.14}_{-0.00}$

Table 2.2: Summary of the theoretical input parameters for non-leptonic B meson decays into two light mesons. All scale-dependent quantities refer to $\mu = 2\,\mathrm{GeV}$ unless otherwise stated.

Part I

Rare B decays in the MSSM with Minimal Flavour Violation

3. The minimally flavour violating MSSM

In this chapter, we present a brief introduction to the Minimal Supersymmetric Standard Model (MSSM) with Minimal Flavour Violation (MFV). In section 3.1 the general set-up of the MSSM is sketched (detailed reviews can e.g. be found in [55]) before we concentrate on the framework of MFV in section 3.2. Here, we start with a detailed discussion of the symmetry-based definition of MFV by D'Ambrosio et al. [7]. Subsequently we define the scenario of "Naive MFV" used in this thesis and compare it to the symmetry-based definition as well as to the framework of "Constrained MFV" (CMFV) defined by Buras et al. [56]. Finally, we elaborate on the low-energy structure of the MSSM with naive MFV pointing out which operators in the effective $\Delta B = \Delta S = 1$ Hamiltonian are expected to receive large contributions.

3.1. Construction of the MSSM

To construct the MSSM one extends the SM by the minimal number of particles needed such that every fermion has a bosonic superpartner and vice versa. The particles are then grouped into two types of supermultiplets: chiral multiplets containing partners of spin 0 and spin 1/2 and vector multiplets containing partners of spin 1/2 and spin 1. Since these supermultiplets transform irreducible under SUSY and the gauge symmetries and since thus the superpartners must have equal gauge quantum numbers, it is not possible to form a pair of superpartners consisting of two SM particles[1]. Therefore one has to add to each SM particle an additional partner particle. Moreover, as we will see, a second Higgs doublet is needed compared to the SM. Names for the new particles are created from the names of their SM partners, in case of spin 0 by putting a s- in front (e.g. top $\to$ stop), in case of spin 1/2 by adding the ending -ino (e.g. Higgs $\to$ Higgsino). The complete particle content of the MSSM is shown in tables 3.1 and 3.2.

Now, after the particle content of the MSSM has been identified, we have to specify in a second step the interactions among the particles. These are completely fixed in SUSY theories by writing down the superpotential, a holomorphic function of the chiral superfields. The superpotential has to be gauge invariant and the theory derived from it should be renormalisable. To avoid interactions which violate lepton or baryon number, an additional symmetry called R-parity has

[1] Grouping the SM Higgs doublet and one of the lepton doublets together into a supermultiplet would violate lepton number.

superfield	spin 0	spin $\frac{1}{2}$	$SU(3)_C \times SU(2)_L \times U(1)_Y$
Q	$\begin{pmatrix} \tilde{u}_L \\ \tilde{d}_L \end{pmatrix}$	$\begin{pmatrix} u_L \\ d_L \end{pmatrix}$	$(3, 2, +\frac{1}{6})$
$\bar{u}$	$\tilde{u}_R^*$	$(u_R)^c$	$(\bar{3}, 1, -\frac{2}{3})$
$\bar{d}$	$\tilde{d}_R^*$	$(d_R)^c$	$(\bar{3}, 1, +\frac{1}{3})$
L	$\begin{pmatrix} \tilde{\nu}_L \\ \tilde{e}_L \end{pmatrix}$	$\begin{pmatrix} \nu_L \\ e_L \end{pmatrix}$	$(1, 2, -\frac{1}{2})$
$\bar{e}$	$\tilde{e}_R^*$	$(e_R)^c$	$(1, 1, +1)$
H_u	$\begin{pmatrix} H_u^+ \\ H_u^0 \end{pmatrix}$	$\begin{pmatrix} \widetilde{H}_u^+ \\ \widetilde{H}_u^0 \end{pmatrix}$	$(1, 2, +\frac{1}{2})$
H_d	$\begin{pmatrix} H_d^0 \\ H_d^- \end{pmatrix}$	$\begin{pmatrix} \widetilde{H}_d^0 \\ \widetilde{H}_d^- \end{pmatrix}$	$(1, 2, -\frac{1}{2})$

Table 3.1: Chiral supermultiplets in the MSSM

to be postulated. The most general superpotential fulfilling all these requirements reads

$$W = y_u^{ij} \bar{u}_i Q_j H_u - y_d^{ij} \bar{d}_i Q_j H_d - y_e^{ij} \bar{e}_i L_j H_d + \mu H_u H_d \,, \tag{3.1}$$

where the product of two $SU(2)_L$ doublets has to be interpreted as $QH \equiv \epsilon_{\alpha\beta} Q^\alpha H^\beta$ with $\epsilon_{\alpha\beta}$ being the two-dimensional antisymmetric tensor. Whereas in the SM the charge-conjugated Higgs field $\widetilde{H} = \epsilon H^*$ is used to generate masses for the up-type quarks, this is not possible in the MSSM because otherwise the superpotential would contain H_d and H_d^* at the same time and would then not be holomorphic anymore. Therefore it is necessary to introduce a second Higgs doublet H_u.

The Yukawa Lagrangian for the quark fields

$$-\mathcal{L}_y = y_u^{ij} \bar{u}_R^i Q_j H_u - y_d^{ij} \bar{d}_R^i Q_j H_d + \text{h.c.} \,, \tag{3.2}$$

which is obtained from the superpotential (3.1) by replacing the (s)quark superfields by their fermionic and the Higgs superfields by their bosonic part, is that of a two-Higgs-doublet model (2HDM) of type II. The Yukawa couplings y_q^{ij} $(q = u, d)$ are arbitrary 3×3 matrices in family space. Allowing different unitary rotations for left-handed up- and down quarks and thus giving

superfield	spin $\frac{1}{2}$	spin 1	$SU(3)_C \times SU(2)_L \times U(1)_Y$
V_1	$\widetilde{B}^0$	B^0	(1,1,0)
V_2	$\widetilde{W}^{\pm}, \widetilde{W}^0$	$W^{\pm}, W^0$	(1,3,0)
V_3	$\tilde{g}$	g	(8,1,0)

Table 3.2: Vector supermultiplets in the MSSM

up manifest $SU(2)$-invariance, it is possible to choose a basis in which the tree-level Yukawa couplings are flavour-diagonal, i.e. $y_q^{ij} = y_{q_j}\delta_{ij}$. In this basis of mass-eigenstates the CKM-matrix enters the couplings of the W-boson to the quark fields. Performing the rotations needed to switch to the quark mass-eigenstates on the whole superfields in order to maintain supersymmetry, one arrives at the Super-CKM basis.

In the course of electroweak symmetry breaking the neutral components of the two Higgs doublets H_u and H_d acquire vacuum expectation values (vevs) v_u and v_d with the sum $v_u^2 + v_d^2$ being fixed to $v^2 \approx (174\ \text{GeV})^2$ and the ratio defined as

$$\tan\beta \equiv v_u/v_d \tag{3.3}$$

remaining as a free parameter. The vevs v_u and v_d give masses to up- and down-type quarks according to

$$m_u = y_u v_u, \qquad m_d = y_d v_d. \tag{3.4}$$

Therefore the parameter $\tan\beta$ determines the ratio of up- and down-type Yukawa couplings which is given in terms of the measured quark masses as

$$\frac{y_d}{y_u} = \frac{m_d}{m_u}\tan\beta. \tag{3.5}$$

From (3.5) we see that a large value of $\tan\beta$ (~ 50) leads to increased down-type Yukawa couplings with y_b of order $\mathcal{O}(1)$ permitting bottom-top Yukawa unification. As it is characteristic for a theory with large dimensionless parameter, various parametric enhancement effects occur which will be studied in the subsequent chapters.

As explained in the outset, SUSY can only be realised in nature as a broken symmetry. It is assumed that, like the electroweak symmetry, SUSY is broken spontaneously by the vev of a scalar field. For phenomenological reasons, this scalar field cannot be one of the MSSM fields itself but has to be part of a so-called hidden sector consisting of additional fields which couple only weakly to the MSSM matter. There exist different theoretical proposals on how the SUSY breaking might be mediated from the hidden sector to the visible MSSM. The most popular ones are based on gravity, gauge interactions or anomalies. However, since the correct mechanism is not known, the simplest approach for phenomenological studies is just to put by hand terms into the Lagrangian which explicitly break SUSY. It is clear that SUSY should be broken in such a way that its attractive features are kept, especially the hierarchy problem should not be restored. SUSY breaking terms fulfilling this request are called "soft" and have been classified for the first time in Ref. [57]. For the MSSM they are given by

$$\begin{aligned} -L_{\text{soft}}^{MSSM} &= \frac{1}{2}\left(M_1\overline{\widetilde{B}}\widetilde{B} + M_2\overline{\widetilde{W}}^a\widetilde{W}^a + M_3\overline{\widetilde{g}}^c\widetilde{g}^c\right) \\ &+ \left(a_u^{ij}\tilde{u}_i^*\widetilde{Q}_jH_u - a_d^{ij}\tilde{d}_i^*\widetilde{Q}_jH_d - a_e^{ij}\tilde{e}_i^*\widetilde{L}_jH_d + bH_uH_d + c.c.\right) \\ &+ \left((\widetilde{m}_Q^2)^{ij}\widetilde{Q}_i^*\widetilde{Q}_j + (\widetilde{m}_u^2)^{ij}\tilde{u}_i^*\tilde{u}_j + (\widetilde{m}_d^2)^{ij}\tilde{d}_i^*\tilde{d}_j + (\widetilde{m}_L^2)^{ij}\widetilde{L}_i^*\widetilde{L}_j + (\widetilde{m}_e^2)^{ij}\tilde{e}_i^*\tilde{e}_j\right) \\ &+ \left(m_{H_u}^2H_u^*H_u + m_{H_d}^2H_d^*H_d\right), \end{aligned} \tag{3.6}$$

with $H_{u,d}$ denoting only the scalar part of the corresponding supermultiplet here. The Lagrangian (3.6) provides mass terms for the gauginos, squarks and Higgs bosons as well as trilinear squark-squark-Higgs couplings and a H_u-H_d mixing term. In its most general form it contains more than 100 free parameters.

3.2. Minimal Flavour Violation

Due to the fact that the soft breaking Lagrangian (3.6) has not been derived from an underlying theory or principle, its parameters are a priori completely arbitrary implying an entirely general flavour structure. On the other hand, when confronted with data on flavour physics, the SUSY breaking terms get highly constrained. Therefore, it is quite natural to assume a structure for these terms which reduces the additional amount of flavour violation to a minimum. Depending on how the word "minimum" is interpreted one can end up with different hypotheses of "Minimal Flavour Violation" and various realisations have been studied in the literature. The most consistent definition has been given in Ref. [7] using a spurion formalism. It will be denoted as MFV in the following and is in a sense the "maximal" version of Minimal Flavour Violation containing other scenarios as approximations or special cases. In this section we will discuss this realisation of MFV in some detail before we define our own framework of naive Minimal Flavour Violation (naive MFV) used in this thesis.

3.2.1. Symmetry-based definition of MFV

To illustrate the idea of the spurion method let us consider for a moment a simple quantum mechanical system, namely an electron in a uniform magnetic field $\vec{B}$. Whereas the Hamiltonian of a free electron is invariant under $SO(3)$ rotations, this symmetry is reduced to $SO(2)$ in presence of the *external* magnetic field since now a direction $\vec{B}/|\vec{B}|$ is distinguished from other directions. On the other hand, when the magnetic field is not treated as external but as part of the system, it will transform under rotations and $SO(3)$ invariance is restored. This point of view has the advantage that now $SO(3)$ invariance can be used to construct the Hamiltonian of the system which is thus dictated to be

$$H = c\vec{S}\vec{B}, \tag{3.7}$$

with c being a constant, $\vec{S}$ the spin of the electron and spatial degrees of freedom ignored[2]. After the Hamiltonian has been set up in this way, the magnetic field can be frozen to its constant value inducing thereby the $SO(3) \to SO(2)$ breaking. At this point the two broken generators can be applied for a last time to choose a convenient coordinate system, e.g. with $\vec{B}$ pointing into the z-direction [58]. By this procedure the three components of the magnetic field are reduced to $B_z = |\vec{B}|$, which is the only physical free parameter of the system.

Let us now transfer this concept to the case of the SM Lagrangian restricting ourselves to the quark sector. The gauge interactions cannot distinguish among the three families of matter and therefore they exhibit a global $U(3)_Q \times U(3)_u \times U(3)_d$ symmetry, where Q denotes the left-handed quark doublet, u and d the right-handed up- and down-quark singlets. This family symmetry is broken by the Yukawa interactions

$$-\mathcal{L}_y^{\rm SM} = \bar{u}_R Y_u Q H + \bar{d}_R Y_d Q \widetilde{H} + \text{h.c.} \tag{3.8}$$

which connect left- and right-handed quark fields via the 3×3 Yukawa matrices Y_u and Y_d. However, in complete analogy to the case of the magnetic field discussed above, we can reinstall the $[U(3)]^3$ symmetry by promoting the Yukawa couplings to dynamical fields (spurions) which transform under the family rotations

$$Q \to R_Q Q, \quad u_R \to R_u u_R, \quad d_R \to R_d d_R \qquad (R_Q \in U(3)_Q, \ldots) \tag{3.9}$$

according to

$$Y_u \to R_u Y_u R_Q^\dagger, \qquad Y_d \to R_d Y_d R_Q^\dagger. \tag{3.10}$$

In the same way as the $SO(3)$ symmetry in the example from quantum mechanics was used to construct the most general Hamiltonian, the restored family symmetry of the SM can be exploited to predict the operators appearing in the effective Hamiltonian (2.1) and to make qualitative statements about their Wilson coefficients. For example, to lowest order in the Yukawa couplings, a FCNC $(V-A)$ - current has to be of the form

$$j^\mu_{ij} \sim \bar{Q}_i (Y_u^\dagger Y_u)_{ij} \gamma^\mu Q_j \approx y_t^2 V_{ti}^* V_{tj} \left(\bar{Q}_i \gamma^\mu Q_j\right) \qquad \text{(no sum over } i \neq j) \tag{3.11}$$

[2] Higher powers of $\vec{S}\vec{B}$ do not appear because $(\vec{B}\vec{S})^2 = \frac{\hbar^2}{4}\vec{B}^2$.

in order to respect the $[U(3)]^3$ symmetry. This is exactly the structure which would result from an explicit loop calculation. Of course, the spurion method is only a technical trick: In the end the Yukawa spurions have to be frozen to their constant values inducing thereby the $[U(3)]^3 \to U(1)_B$ breaking with the remaining $U(1)$ symmetry being baryon number. Analogously to the example of the magnetic field, we can use the $3 \cdot 9 - 1 = 26$ broken generators to choose a convenient basis in family space [58]. By this procedure the 36 real parameters of the two complex 3×3 Yukawa matrices reduce to 10 physical ones: 6 diagonal couplings in direct correspondence to the quark masses and 4 parameters constituting the CKM matrix, namely three mixing angles and one phase.

A generic extension of the SM will typically introduce further $[U(3)]^3$-breaking interactions. To restore the family symmetry one has therefore in general to promote additional couplings to spurions. This observation suggests the following definition of MFV [7]: An extension of the SM is called minimally flavour violating, if the Yukawa spurions are the only spurions needed to re-establish the $[U(3)]^3$ family symmetry and if every interaction term in the Lagrangian can be expressed *in a natural way* in terms of the Yukawa spurions. In contrast to most other formulations of Minimal Flavour Violation, this definition can be applied to any new physics extension of the SM, not only to the MSSM. It ensures that FCNC transitions are still governed by expressions like (3.11), i.e. controlled by the hierarchical structure of the CKM matrix *and* the Yukawa couplings so that the same suppression mechanisms as in the SM take effect.

To apply the MFV hypothesis to the MSSM we parametrise the matrices $\widetilde{m}_Q^2$, $\widetilde{m}_u^2$, $\widetilde{m}_d^2$, a_u and a_d in terms of the Yukawa spurions such that the soft SUSY breaking Lagrangian (3.6) becomes $[U(3)]^3$-invariant [7]:

$$
\begin{aligned}
\widetilde{m}_Q^2 &= m_0^2 \left[a_1 + b_1\, Y_u^\dagger Y_u + b_2\, Y_d^\dagger Y_d + (b_3\, Y_d^\dagger Y_d Y_u^\dagger Y_u + \text{h.c.}) + \ldots\right], \\
\widetilde{m}_u^2 &= m_0^2 \left[a_2 + b_5\, Y_u Y_u^\dagger + \ldots\right], \\
\widetilde{m}_D^2 &= m_0^2 \left[a_3 + b_6\, Y_d Y_d^\dagger + \ldots\right], \\
a_u &= A_0\, Y_u \left(a_4 + b_7\, Y_d^\dagger Y_d + \ldots\right), \\
a_d &= A_0\, Y_d \left(a_5 + b_8\, Y_u^\dagger Y_u + \ldots\right).
\end{aligned} \tag{3.12}
$$

Adopting vector space language, equation (3.12) displays the projection of the soft SUSY breaking matrices onto the subspace spanned by appropriate combinations of the Yukawa spurions. However, for example in the case of $\widetilde{m}_Q^2$ it can be shown that these combinations form a complete basis for the vector space of hermitian 3×3 matrices [59]. Thus from (3.12) alone, no conclusion can be drawn on the structure of $\widetilde{m}_Q^2$. On the other hand, due to the hierarchical structure of the Yukawa matrices, the basis vectors are almost aligned. Therefore, if $\widetilde{m}_Q^2$ would be a completely generic hermitian matrix, some of its coefficients $a_i, b_i, \ldots$ would have to be orders of magnitudes larger than others in contradiction to the naturality of the Yukawa decomposition required by the definition of MFV. Demanding in contrast $a_i, b_i, ... = \mathcal{O}(1)$ allows to truncate the Yukawa decomposition after the first few terms as has been done in (3.12) and results in highly non generic structures for the soft breaking terms. The free parameters in the SUSY breaking sector are then reduced to the $\mathcal{O}(1)$ coefficients $a_i, b_i, ...$ on which this definition of MFV makes no further assumptions.

3.2.2. Naive MFV

Naively one might think that it should be possible to construct an even "more minimal" realisation of MFV by simply setting all the b_i in (3.12) to zero. Then the SUSY breaking terms would be flavour-diagonal in a Super-CKM basis which is constructed by rotating the superfields in such a way that the up- and down-type Yukawa couplings are simultaneously diagonalised. As a consequence, all gluino-squark-quark and neutralino-squark-quark couplings in the MSSM Lagrangian would be flavour-conserving. Further the chargino-squark-quark couplings would come with the same CKM elements as the corresponding couplings of W bosons or charged Higgs bosons to (s)quarks. One would expect such a scenario of naive Minimal Flavour Violation (naive MFV) to occur if SUSY is broken by a flavour-blind mechanism leading to flavour-universal squark mass matrices.

However, it turns out that the definition of naive MFV is not renormalisation group invariant: Eliminated at a certain scale, the b_i terms reappear at any other scale, generated by the RGE running of the soft breaking terms. Therefore, even if one imposes flavour universality at the SUSY breaking scale, the assumption of naive MFV at the low energy electroweak scale can be regarded only as an approximation which in the first instance is expected to work well only if the SUSY breaking scale is not too high. However, in our version of naive MFV we slightly go beyond flavour universality, as we allow the SUSY-breaking terms of the third generation to be different from those of the first two[3]. In this way we account for RGE running involving the large top and bottom Yukawa couplings and thus include also the cases of the widely-studied CMSSM (see e.g. Refs. [8] for recent studies) and mSUGRA [9] models, in which the universal boundary condition is imposed at the GUT scale. Moreover, even though in models with high-scale flavour universality the RGE induces flavour-violating gluino and neutralino couplings at the electroweak scale, their impact on FCNC transitions like $B-\overline{B}$ mixing and $b \to s\gamma$ is small [60] and so the naive MFV pattern essentially stays intact.

It should be stressed that our scenario of naive MFV differs from the one called "Constrained MFV" (CMFV) defined by Buras et al. [56]. The definition of CMFV requires the structure of

[3]It should be stressed that this is possible for the right-handed bilinear mass terms but not for the left-handed ones: In the Super-CKM basis one has $m^2_{\tilde{d}_L} = \widetilde{m}^2_Q$ and $m^2_{\tilde{u}_L} = V\widetilde{m}^2_Q V^\dagger$. The naive MFV hypothesis of diagonal $\widetilde{m}^2_{d_L}, \widetilde{m}^2_{u_L}$ matrices therefore implies $\widetilde{m}^2_{u_L,d_L} \propto \mathbb{1}$.

low-energy operators to be the same as in the SM. As we will see in the next section, this is not the case for the MSSM with naive MFV.

Finally, we emphasise that no variant of the MFV assumption forbids flavour-diagonal CP-violating phases [24]. Such phases appear in trilinear terms A_i, the higgsino mass parameter μ, and the gaugino mass terms M_i, $i = 1, 2, 3$, which are consequently treated as complex quantities throughout this thesis. Only certain phase differences are physical, CP-violating quantities. We choose a phase convention in which the gluino mass parameter M_3 is real and positive, so that $M_3 = m_{\tilde{g}}$.

For $M_{\text{SUSY}} \sim v$, the Higgs fields H_u and H_d induce sizeable mixing between the left- and right-handed squarks $\tilde{q}_L$ and $\tilde{q}_R$, which causes the mass eigenstates $\tilde{q}_{1,2}$ to be different from $\tilde{q}_{L,R}$. The charged winos and higgsinos mix as well, forming chargino mass eigenstates $\widetilde{\chi}^{\pm}_{1,2}$, and so do the bino, the neutral wino and the neutral higgsinos, forming neutralino mass eigenstates $\widetilde{\chi}^{0}_{1..4}$. Our conventions concerning the mass and mixing matrices can be found in Appendix A.1. We always work in the Super-CKM basis and use the conventions of the SUSY Les Houches Accord (SLHA) [61]. The phases entering the left-right mixing of squarks are unspecified by the SLHA and are defined in Appendix A.1.

3.2.3. Low-scale structure of the MSSM with MFV

The spurion method developed in section 3.2.1 can be used to construct the low-energy effective Hamiltonian of a theory respecting the MFV hypothesis [7]. Neglecting all Yukawa couplings except for y_t, the only relevant non-trivial flavour structure is given by

$$(Y_u^\dagger Y_u)_{ij} \approx \frac{m_t^2}{v^2} V_{ti}^* V_{tj}. \tag{3.13}$$

If one considers all $[U(3)]^3$-invariant $\Delta B = \Delta S = 1$ - operators of dimension 6 and applies (3.13) to them, one ends up with the effective Hamiltonian (2.1) of the SM. Since the Yukawa spurions are responsible for the operators appearing in the effective Hamiltonian, one might expect MFV models to lead to the same low energy structure (2.1) as the SM. However, there is an additional flavour pattern in the MSSM with MFV originating from the fact, that large values of $\tan\beta$ enhance the down-type Yukawa couplings (see (3.5)) rendering

$$(Y_d)_{ij} \sim \frac{m_{d_i}}{v \cos\beta} \delta_{ij} \tag{3.14}$$

non-negligible. The consequences on the operator basis of the $\Delta B = \Delta S = 1$ - Hamiltonian are as follows:

- The magnetic and chromomagnetic operators $Q_{7\gamma,8g}$ emerge from the flavour structure

$$\bar{Q}_i\,(Y_u^\dagger Y_u Y_d^\dagger)_{ij}\,\sigma^{\mu\nu} d_R^j \;\to\; y_b\, y_t^2\, V_{ts}^* V_{tb}\,(s_L\sigma^{\mu\nu} b_R). \tag{3.15}$$

Since this operator violates $SU(2)_L$ invariance, its coefficient has to be proportional to an electroweak vev. For the SM contribution this vev is generated by attaching the Higgs field H_d to the external b quark line. In the MSSM, one can instead couple the Higgs field H_u to a SUSY particle running in the FCNC loop. In this case the corresponding contribution to $C_{7\gamma,8g}$ is $\tan\beta$-enhanced compared to the SM one because of

$$H_d \to v_d, \qquad\qquad H_u \to v_u = v_d \tan\beta. \tag{3.16}$$

- The potential enhancement of down-type (and lepton) Yukawa couplings allows to construct also scalar "penguin" operators

$$\left(\bar{Q}_i\,(Y_u^\dagger Y_u Y_d^\dagger)_{ij}\, d_R^j\right)\,(\bar{\psi}_R^m\,(Y_\psi)_{mn}\,\Psi_n) \;\to\; y_b\, y_{\psi_m}\, y_t^2\, V_{ts}^* V_{tb}\,(s_L b_R)(\bar{\psi}_R^m \psi_L^m), \tag{3.17}$$
$$\left(\bar{Q}_i\,(Y_u^\dagger Y_u Y_d^\dagger)_{ij}\, d_R^j\right)\left(\bar{\Psi}_m\,(Y_\psi^\dagger)_{mn}\,\psi_R^n\right) \;\to\; y_b\, y_{\psi_m}\, y_t^2\, V_{ts}^* V_{tb}\,(s_L b_R)(\bar{\psi}_L^m \psi_R^m) \tag{3.18}$$

with $\Psi_m = Q_m, L_m$, $\psi_R^m = d_R^m, \ell_R^m$ and $\psi_L^m = d_L^m, \ell_L^m$. The Wilson coefficient of the operator (3.17) receives contributions from neutral-Higgs penguin diagrams which can play in the league of tree-level contributions thanks to a chiral enhancement. This property makes $B_s \to \mu^+\mu^-$ the standard candle for the large-$\tan\beta$ region of the MSSM. The coresponding four-quark operators have less significance because the $\psi = b$ operator contributes to rare B decays only at NLO in the effective theory or via small RGE evolution effects and the $\psi = s$ operator suffers from a m_s/m_b suppression compared to the former one. Therefore correlation with the $\psi = \ell$ operator and the present upper limit on $B_s \to \mu^+\mu^-$ [19] render effects of the four-quark operators negligible [49]. Note that similar neutral-Higgs contributions do not occur in the case of the second operator (3.18). This is a result of a Peccei-Quinn (PQ) symmetry

$$\begin{aligned} &\psi_R \to e^{i\delta}\psi_R, \qquad && u_R \to u_R, \qquad && \Psi \to \Psi, \\ &H_d \to e^{i\delta} H_d, && H_u \to H_u, && \end{aligned} \tag{3.19}$$

obeyed by the tree-level Yukawa Lagrangian and the tree-level Higgs potential of the MSSM. Since the operator (3.18) breaks this symmetry (and in addition $SU(2)_L$ invariance), its coefficient is proportional to a small PQ-breaking parameter in the Higgs potential [62] (and in addition to a factor $v^2/M_{A^0}^2$).

- In a similar way, tree-level charged-Higgs exchange generates scalar "current-current" operators which read

$$\left(\bar{Q}_i\,(Y_d^\dagger)_{ij}\,d_R^j\right)\,(\bar{\varphi}_R^m\,(Y_\varphi)_{mn}\,\Phi_n) \;\rightarrow\; y_b\,y_{\varphi_m}\,V_{cb}\,(c_L b_R)(\varphi_R^m \phi_L^m), \tag{3.20}$$

$$\left(\bar{Q}_i\,(Y_d^\dagger)_{ij}\,d_R^j\right)\,(\bar{\Phi}_m\,(Y_\varphi^\dagger)_{mn}\,\varphi_R^n) \;\rightarrow\; y_b\,y_{\varphi_m}\,V_{cb}\,(c_L b_R)(\phi_L^m \varphi_R^m) \tag{3.21}$$

with $\Phi_m = Q_m, L_m$, $\varphi_R^m = d_R^m, \ell_R^m$ and $\phi_L^m = u_L^m, \nu_L^m$. As in the neutral-Higgs counterpart, the operator in the second line is suppressed by a small PQ-breaking parameter. The first operator with $\varphi = \tau$, on the other hand, is responsible for large effects in the decays $B^+ \to \tau^+\nu$ [15] and $B^+ \to D\tau^+\nu$ [16]. Measurements of these decays limit corresponding effects in the $\varphi = c$ four-quark operator whose coefficient is moreover suppressed compared to the $\varphi = \tau$ one by m_s/m_τ.

Focussing on non-leptonic B decays, we note that no additional operators compared to the SM have to be considered since effects of the scalar operators are numerically small. SUSY contributions to the SM operators decouple with v^2/M_{SUSY}^2, can in case of $Q_{7\gamma}$ and Q_{8g} however be enhanced by a factor of $\tan\beta$.

4. Resummation of $\tan\beta$ - enhanced loop corrections beyond the decoupling limit

Large values ~ 50 of the dimensionless parameter $\tan\beta$, introduced in the last chapter, lead to a parametric enhancement of certain loop corrections to Feynman amplitudes. These contributions corrupt a naive expansion of Feynman amplitudes in the loop order. A fixed-order calculation has to be supported by a resummation of such $\tan\beta$-enhanced effects to all orders.

In this chapter we derive resummation formulae for $\tan\beta$-enhanced loop corrections which are valid for arbitrary values of the SUSY mass parameters. Several of these formulae were previously known only in the limit $M_{\text{SUSY}} \gg v, M_{A^0,H^0,H^\pm}$ which permits application of an effective theory approach. After a brief discussion of this method in Section 4.1 we oppose to it our diagrammatic approach in Section 4.2. This approach is then applied to the flavour-conserving case in Section 4.3, where we clarify the scheme dependence of the resulting resummation formula, and for the first time extended to the flavour-changing case, where we present in Sections 4.4 and 4.5 two different procedures for the resummation. In Section 4.5.2 the resummed $\tan\beta$-enhanced effects are finally cast into a set of effective Feynman rules which allow for an easy implementation into automatic calculations.

4.1. Effective theory approach for $M_{\text{SUSY}} \gg v, M_{A^0,H^0,H^\pm}$

According to the Lagrangian (3.2), the right-handed down quarks d_R^i couple at tree level only to the Higgs field H_d but not to H_u. This is a consequence of the requirement that the superpotential has to be a holomorphic function of the fields. On the other hand, right-handed down quarks can couple to H_u via loops of SUSY-particles, e.g. of squarks and gluinos, which we assume for the moment to be much heavier than the Higgs fields A^0, H^0, $H^\pm$ and the SM particles. Under this assumption the SUSY-particles can be integrated out and this procedure leads to an effective non-holomorphic coupling $\widetilde{y}_d^{ij}$ of H_u to d_R^i [63] (Fig. 4.1). In the Super-CKM-basis for the quark and squark fields, in which $y_d^{ij} = y_{d_i}\delta^{ij}$, the Yukawa couplings of the down-type quarks are then given by the effective Yukawa Lagrangian

$$-\mathcal{L}_{y,d}^{\text{eff}} = -y_{d_i}\bar{d}_R^i Q_i H_d + \widetilde{y}_d^{ij}\bar{d}_R^i Q_j H_u^* + \text{h.c.} \tag{4.1}$$

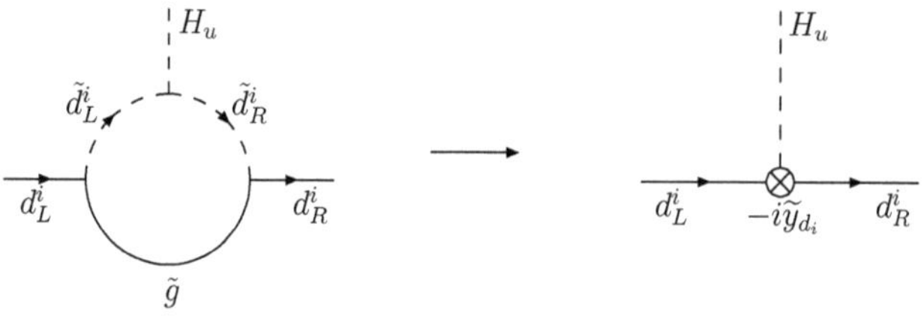

Figure 4.1: Effective coupling of the down-type quarks to H_u

The dominant contribution to the effective coupling $\widetilde{y}_d^{ij}$ stems from a gluino-squark-loop and is depicted in Fig. 4.1. In our framework of naive MFV, it is flavour-diagonal with

$$\widetilde{y}^{\tilde{g}}_{d_i} = y_{d_i}\, \epsilon^{\tilde{g}}_i(\mu, m_{\tilde{d}^i_L}, m_{\tilde{d}^i_R}), \tag{4.2}$$

and

$$\epsilon^{\tilde{g}}_i(\mu, m_{\tilde{d}^i_L}, m_{\tilde{d}^i_R}) = -\frac{2\alpha_s}{3\pi} m_{\tilde{g}} \mu^*\, C_0(m_{\tilde{g}}, m_{\tilde{d}^i_L}, m_{\tilde{d}^i_R}). \tag{4.3}$$

Here $m^2_{\tilde{d}^i_L}$ and $m^2_{\tilde{d}^i_R}$ are the mass terms for the left- and right-handed down-squarks of the i-th generation, respectively, $m_{\tilde{g}}$ is the gluino mass and the loop integral C_0 is defined in Appendix A.2. Accounting for similar contributions from loops with charginos (still neglecting flavour mixing) or neutralinos we write $\epsilon_i = \epsilon^{\tilde{g}}_i + \epsilon^{\tilde{\chi}^\pm}_i + \epsilon^{\tilde{\chi}^0}_i$.

When replaced by their vevs, both Higgs fields H_u and H_d generate mass terms $m^{(u)}_{d_i}$ and $m^{(d)}_{d_i}$ for the down-type quarks via the Yukawa couplings in $\mathcal{L}^{\text{eff}}_{y,d}$. The ratio of these mass terms is given by

$$\Delta_i \equiv \frac{m^{(u)}_{d_i}}{m^{(d)}_{d_i}} = \frac{\widetilde{y}_{d_i} v_u}{y_{d_i} v_d} = \epsilon_i \tan\beta. \tag{4.4}$$

A large value of $\tan\beta$ can compensate for the loop factor ϵ_i rendering $m^{(u)}_{d_i}$ and $m^{(d)}_{d_i}$ of the same order of magnitude. Treating them therefore on an equal footing, we get a modified relation between the Yukawa coupling y_{d_i} and the physical quark mass m_{d_i}:

$$m_{d_i} = m^{(d)}_{d_i} + m^{(u)}_{d_i} \quad \Rightarrow \quad y_{d_i} = \frac{m_{d_i}}{v_d(1+\epsilon_i \tan\beta)}. \tag{4.5}$$

In the effective theory approach, the $\tan\beta$-enhanced self-energy contributions (e.g. $m^{(u)}_{d_i}$ in the previous flavour-conserving example) are generated by the operators $Q_{ij} = \bar{d}^i_R Q_j H^*_u$. From this observation, we can immediately conclude that the resulting contributions have the following properties:

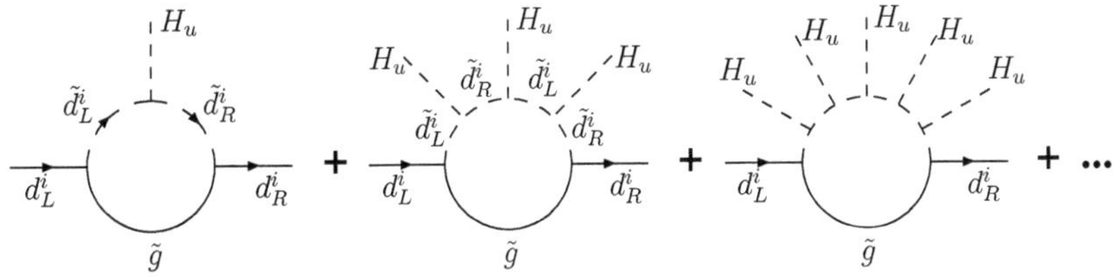

Figure 4.2: Series of 'hedgehog diagrams' contributing to m_{d_i}

- They do not decouple for $M_{\rm SUSY} \to \infty$, because the operators Q_{ij} have dimension four and hence the couplings $\widetilde{y}_d^{ij}$ are dimensionless.
- They are finite. Otherwise appropriate counterterms $\delta\widetilde{y}_d^{ij}$ would be needed to cancel divergences. However, such counterterms are not allowed, because the non-holomorphic couplings $\widetilde{y}_d^{ij}$ are absent at tree-level.
- There are no genuine higher loop contributions of the form $(\epsilon \tan\beta)^n$ with $n \geq 2$. This can be seen by recognizing that the $\tan\beta$ - enhancement in the ratio Δ_i stems from the $\cos\beta$-suppression of $m_{d_i}^{(d)} = y_{d_i} v \cos\beta$. Therefore only one factor of $\tan\beta$ appears, independently of the number of loops and the number of powers of $v_u = v\sin\beta$ considered in $m_{d_i}^{(u)}$.

It is illustrative to consider the extension of the effective-field-theory formalism to subleading powers in $v^2/M_{\rm SUSY}^2$. Apart from the Q_{ij} one must then add also operators to $\mathcal{L}_{y,d}^{\rm eff}$ which involve a higher number $k \geq 3$ of H_u fields. Contributions to these higher-dimensional operators are suppressed for $M_{\rm SUSY} \gg v$ but have to be summed to all orders in k for $M_{\rm SUSY} \sim v$. The gluino contributions to these new effective couplings are shown in Fig. 4.2. In this simple case one can sum the contributions of these 'hedgehog diagrams' to m_{d_i} to all orders in k. The effect of this Dyson-like summation can be incorporated into (4.5) by the replacement

$$\epsilon_i(\mu, m_{\tilde{d}_L^i}, m_{\tilde{d}_R^i}) \to \epsilon_i(\mu, m_{\tilde{d}_1^i}, m_{\tilde{d}_2^i}), \qquad (4.6)$$

where $m_{\tilde{d}_{1,2}^i}$ denote the physical squark masses, i.e. the eigenvalues of the squark mass matrix. Using this expression in Eqs. (4.4) and (4.5) reproduces the result of the diagrammatic resummation of Ref. [14].

In the next sections we will derive resummation formulae for $\tan\beta$-enhanced corrections which are valid for any value of $M_{\rm SUSY}$. We will use and extent the diagrammatic method of Ref. [14] which has the following advantages compared to the effective-field-theory approach:

- Working in the mass-eigenbasis for the squarks, we automatically include contributions of all orders in $v^2/M_{\rm SUSY}^2$ without the need for a summation of a Dyson-like series of diagrams.

- Even though it has been determined by a one-loop calculation, Eq. (4.5) contains contributions $(\epsilon_i \tan\beta)^n$ to all orders in $n = 1, 2, ...$ according to

$$\frac{m_{d_i}}{v_d(1+\epsilon_i \tan\beta)} = \frac{m_{d_i}}{v_d} \sum_n (-\epsilon_i \tan\beta)^n. \tag{4.7}$$

  The origin of the terms of higher loop-order will become transparent in the diagrammatic approach where an explicit resummation is performed.

- The diagrammatic method provides full control over the renormalisation scheme for the SUSY input. This is important because the resulting resummation formulae depend on the scheme which is used. It turns out that they look different also for schemes which differ only by terms of order $v/M_{\rm SUSY}$. Therefore it is difficult to achieve control over the renormalisation scheme in the effective-field-theory approach, even in cases where one succeeds in resumming higher orders in $v^2/M_{\rm SUSY}^2$ as in (4.6).

- We will study $\tan\beta$-enhanced effects in the couplings of SUSY particles like the quark-squark-gluino vertex. The resulting formulae can be applied to low energy processes involving virtual SUSY particles as well as to high energy processes with external SUSY particles. Such an analysis cannot be carried out consistently in the effective-field-theory approach with the SUSY particles integrated out.

4.2. Diagrammatic resummation

Given the interaction Lagrangian $\mathcal{L}$ of a quantum field theory, the transition amplitude $\mathcal{M}$ for a scattering process involving $k \to n$ particles is calculated according to

$$i\mathcal{M} = \frac{\langle 0|\, T\left\{\phi_1(x_1)...\phi_k(x_k)\, \psi_1(y_1)...\psi_n(y_n)\, \exp\left(i \int d^4x\, \mathcal{L}\right)\right\}\, |0\rangle}{\langle 0|\, T\left\{\exp\left(i \int d^4x\, \mathcal{L}\right)\right\} |0\rangle}. \tag{4.8}$$

Here T denotes the time-ordered product. The fields $\phi_1, ..., \phi_k, \psi_1, ..., \psi_n$ representing the incomming and outgoing particles, respectively, are given in the interaction picture. For perturbative calculations, the exponential is expanded to a fixed order in powers of $\mathcal{L}$ corresponding to a fixed loop-order in the graphical representation of (4.8) in terms of Feynman diagrams. The expansion parameters of the the perturbative series are small loop factors $\epsilon \sim \kappa^2/(16\pi^2)$ with $\kappa \lesssim 1$ being coupling constants of the interaction terms in $\mathcal{L}$. However, as demonstrated in the last section, the

ϵ-suppression of superficially subleading contributions can be lifted in the MSSM by an enhancing factor of $\tan\beta$. Therefore, a perturbative calculation to the kth order should rather include all corrections of the form $\epsilon^k(\epsilon\tan\beta)^n$ for $n = 1, 2, ...$. We will now develop a framework which allows to take into account such effects systematically.

To this end we decompose the Lagrangian in the usual way as

$$\mathcal{L} = \mathcal{L}_{\text{ren}} + \mathcal{L}_{\text{ct}} \tag{4.9}$$

where $\mathcal{L}_{\text{ren}}$ is obtained from $\mathcal{L}$ by replacing bare quantities by renormalised ones and $\mathcal{L}_{\text{ct}}$ contains counterterms needed to cancel divergences appearing in loop calculations. The renormalised parameters are "pseudo-observables" [1] which can (at least in principle) be determined from experiment. The decision which physical quantities to use as these "pseudo-observables" defines a renormalisation scheme and fixes the finite part of the counterterms in $\mathcal{L}_{\text{ct}}$. Depending on the renormalisation scheme, finite loop effects at a certain energy scale are absorbed into $\mathcal{L}_{\text{ct}}$. For this reason, $\mathcal{L}_{\text{ct}}$ is a further potential source of $\tan\beta$-enhanced corrections in addition to loop-corrections encountered in the calculation of $\mathcal{M}$ from $\mathcal{L}_{\text{ren}}$. In this section we will identify all types of $\tan\beta$-enhancement effects in $\mathcal{M}$ and $\mathcal{L}_{\text{ct}}$ postponing the discussion of their resummation to the subsequent sections.

4.2.1. $\tan\beta$-enhancement in $\mathcal{M}$

In order to identify $\tan\beta$-enhanced corrections to a given transition matrix element $\mathcal{M}$ we must distinguish two cases, namely whether the leading order contribution $\mathcal{M}_{\text{LO}}$ does involve a $\cos\beta$-suppression or not. In the case of unsuppressed $\mathcal{M}_{\text{LO}}$ a loop correction can only be $\tan\beta$-enhanced if it involves at least one inverse power of $m_b \sim \cos\beta$ [2]. The presence of such inverse powers of m_b is related to the infrared behaviour of $\mathcal{M}$ for $m_b \to 0$. The Kinoshita-Lee-Nauenberg theorem [64] guarantees the absence of power-like divergences for $m_b \to 0$ in genuine multi-loop diagrams. Therefore, $\tan\beta$-enhanced loop corrections can only be generated by insertion of self-energy subdiagrams into quark lines, typically flavour-changing self-energies into external quark legs. This type of $\tan\beta$-enhanced corrections will be studied in Section 4.4.

In case of an explicit $\cos\beta$-suppression of $\mathcal{M}_{\text{LO}}$ the situation is different. Any generic loop correction which does not suffer from the same suppression is $\tan\beta$-enhanced with respect to $\mathcal{M}_{\text{LO}}$. Since this type of $\tan\beta$-enhancement does not replicate itself in higher orders, it can be taken into account by a NLO calculation and no resummation is needed. Examples are the h^0 coupling to down-type quarks and the H^+ coupling to left-handed down-type quarks. The $\tan\beta$-enhanced vertex corrections to these couplings have been studied in Refs. [21, 22].

[1] couplings are, of course, no direct observables but can be determined from cross section measurements.
[2] we neglect the d- and s-quark masses

4.2.2. tan β-enhancement in $\mathcal{L}_{\text{ct}}$

In $\mathcal{L}_{\text{ct}}$ the quark mass counterterm δm_b (or equivalently the Yukawa counterterm δy_b) as well as the CKM counterterms δV_{ij} are sources of $\tan\beta$-enhancement. Of course, these counterterms depend on the chosen renormalisation scheme. In order to be able to use numerical values for m_b and V_{ij} determined from low-energy experiments, we use an appropriate decoupling scheme applying an on-shell subtraction to the $\tan\beta$-enhanced SUSY loops. The resulting one-loop counterterms δy_b and δV_{ij} are of the form $\epsilon\tan\beta$ with ϵ being the respective loop factor. Reinsertion of these counterterms into $\tan\beta$-enhanced loop diagrams gives enhanced higher order contributions to be subtracted by higher order counterterms. These effects can be taken into account by resumming δy_b and δV_{ij} to all orders $(\epsilon\tan\beta)^n$ $(n = 1, 2, ...)$. The corresponding resummation is performed in Section 4.3.2 for δy_b and in Section 4.4.4 for the δV_{ij}.

All $\tan\beta$-enhanced counterterms are finite and so they are absent if a minimal subtraction scheme is chosen for m_b and V_{ij}. However, for example in the case of the bottom mass, the input value $\widetilde{m}_b$ in such a scheme is obtained from the measured $\overline{\text{MS}}$ mass m_b by adding the $\tan\beta$-enhanced self-energy $\Sigma^{RL}_{b,\,\text{SUSY}}(\widetilde{m}_b)$. Therefore, $\widetilde{m}_b$ implicitly contains $\tan\beta$-enhanced corrections and the issue of their resummation would have to be addressed in the determination of a conversion formula between $\widetilde{m}_b$ and m_b.

The prescription of an on-shell subtraction for the SUSY loops contributing to m_b and V_{ij} defines only one part of the total renormalisation scheme. Two such schemes can for example still differ by the renormalisation conditions imposed on the SUSY breaking parameters. As we will see in Section 4.3.3, the resulting resummation formula for δy_b depends on the renormalisation prescriptions applied to the down squark sector.

Finally, we remark that the $\tan\beta$-enhanced self-energy insertions into external quark legs which have been mentioned in the first paragraph of the last section can be absorbed into wave-function counterterms δZ^L and δZ^R of left- and right-handed down-quark fields. These counterterms $\delta Z^{L,R}$ are 3×3 matrices in flavour space and their determination and resummation is discussed in Section 4.5.1. The consideration of $\delta Z^{L,R}_{ij}$-insertions into leading order diagrams supersedes then an explicit calculation of the corresponding external leg corrections to $\mathcal{M}_{\text{LO}}$. In this way, this type of $\tan\beta$-enhanced corrections moves from the category of transition amplitude effects to the category of counterterm effects.

The $\tan\beta$-enhanced counterterm effects can easily be incorporated into an automated leading order calculation of Feynman diagrams. To this end one has just to treat enhanced counterterm vertices on an equal footing with tree-level vertices by formulating effective Feynman rules. These Feynman rules, which can easily be implemented into computer programs like FeynArts [28], are formulated in Section 4.5.2. On the other hand, in order to include the $\tan\beta$ - enhanced effects discussed in the last section, one has to single the enhanced diagrams out of the bulk of NLO corrections and calculate them explicitly. This procedure is not well suited for an automated implementation into computer programs like FeynArts which are designed to perform fixed-order calculations. In this context, it is favourable to replace $\tan\beta$-enhanced external leg corrections by matrix-valued wave-function counterterms as proposed in the last paragraph. Then also these effects can be cast into effective Feynman rules and they are included in the rules given in Section 4.5.2. The only $\tan\beta$-enhanced corrections which can not be incorporated by means of effective Feynman rules, are then the corrections to $\cos\beta$-suppressed amplitudes $\mathcal{M}_{\text{LO}}$. However, such corrections arise only for a few number of amplitudes and to take them into account a full NLO calculation is required anyway. Therefore, we will not concern ourselves with this type of $\tan\beta$-enhancement any further. Instead we will discuss all the other enhanced effects and their resummation now in detail.

4.3. The flavour-conserving case

As demonstrated in Section 4.1, effective non-holomorphic couplings of right-handed down-type quarks to H_u lead to self-energy contributions to their masses which are $\tan\beta$-enhanced with respect to the tree level values. In the diagrammatic approach of Section 4.2 these $\tan\beta$-enhanced self-energies renormalise the masses m_{d_i} and Yukawa couplings y_{d_i}. In this section we determine the resummed counterterm δy_{d_i} and discuss its scheme dependence. For definiteness we quote the results for the b-quark. Expressions for d- and s-quarks are obtained by obvious replacements.

4.3.1. Flavour-conserving $\tan\beta$-enhanced self-energies

A general contribution to a fermion self-energy can be decomposed as

$$\begin{aligned}\Sigma_b(p) = \not{p}\left[\Sigma_b^{LL}(p^2)P_L + \Sigma_b^{RR}(p^2)P_R\right] \; + \; \Sigma_b^{RL}(p^2)P_L \; + \; \Sigma_b^{LR}(p^2)P_R \\ \text{with} \quad \Sigma_b^{LR}(p^2) = \left(\Sigma_b^{RL}(p^2)\right)^*,\end{aligned} \tag{4.10}$$

where p is the external momentum. The scalar part Σ_b^{RL} contains $\tan\beta$-enhanced pieces from gluino-squark, chargino-squark and neutralino-squark loops. These contributions are depicted in Fig. 4.3 and read

$$\Sigma_b^{RL} \;=\; m_b\Delta_b \qquad \text{with} \qquad \Delta_b \;=\; \Delta_b^{\tilde{g}} + \Delta_b^{\tilde{\chi}^\pm} + \Delta_b^{\tilde{\chi}^0} \tag{4.11}$$

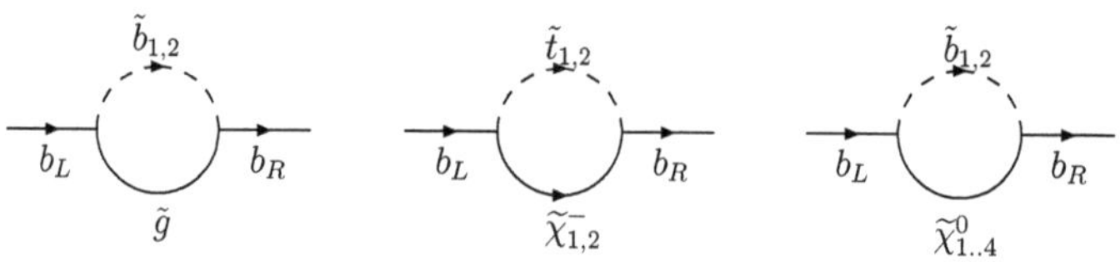

Figure 4.3: $\tan\beta$-enhanced self-energy diagrams with (from left to right) gluinos, charginos and neutralinos.

and

$$\Delta_b^{\tilde{g}} = \frac{\alpha_s}{3\pi}\frac{m_{\tilde{g}}}{m_b}\sin 2\tilde{\theta}_b\, e^{-i\tilde{\phi}_b}\cdot\left[B_0(m_{\tilde{g}},m_{\tilde{b}_1}) - B_0(m_{\tilde{g}},m_{\tilde{b}_2})\right], \tag{4.12}$$

$$\begin{aligned}\Delta_b^{\tilde{\chi}^\pm} = & -\frac{g^2}{16\pi^2}\frac{1}{\cos\beta}\sum_{m=1}^{2}\left\{\frac{m_{\tilde{\chi}^\pm_m}}{2\sqrt{2}M_W}\frac{y_t}{g}\widetilde{U}^*_{m2}\widetilde{V}^*_{m2}\sin 2\tilde{\theta}_t e^{i\tilde{\phi}_t}\right.\\ &\cdot\left[B_0(m_{\tilde{\chi}^\pm_m},m_{\tilde{t}_1}) - B_0(m_{\tilde{\chi}^\pm_m},m_{\tilde{t}_2})\right]\\ &\left.-\frac{m_{\tilde{\chi}^\pm_m}}{\sqrt{2}M_W}\widetilde{U}^*_{m2}\widetilde{V}^*_{m1}\left[\cos^2\tilde{\theta}_t\, B_0(m_{\tilde{\chi}^\pm_m},m_{\tilde{t}_1}) + \sin^2\tilde{\theta}_t\, B_0(m_{\tilde{\chi}^\pm_m},m_{\tilde{t}_2})\right]\right\}, \end{aligned}\tag{4.13}$$

$$\begin{aligned}\Delta_b^{\tilde{\chi}^0} = & \frac{g^2}{16\pi^2}\frac{1}{\cos\beta}\sum_{m=1}^{4}\frac{m_{\tilde{\chi}^0_m}}{2M_W}\widetilde{N}^*_{m2}\widetilde{N}^*_{m3}\\ &\cdot\left[\cos^2\tilde{\theta}_b\, B_0(m_{\tilde{\chi}^0_m},m_{\tilde{b}_1}) + \sin^2\tilde{\theta}_b\, B_0(m_{\tilde{\chi}^0_m},m_{\tilde{b}_2})\right]. \end{aligned}\tag{4.14}$$

By evaluating Σ_b^{RL} for $p^2 = 0$, we have neglected terms suppressed by $m_b^2/M_{\rm SUSY}^2$ and terms suppressed by $\cot\beta$. In the effective theory approach, $\cot\beta$-suppressed terms are generated by the Higgs field H_d which can couple either to the sparticle loop or (in contrast to H_u) also to the external down-quark line. The coupling to the external quark line corresponds in the diagrammatic language to a chirality flip obtained by application of the Dirac equation $\not{p}\, b_{L,R} = m_b\, b_{R,L}$. Such contributions are omitted if the self-energy is evaluated for vanishing external momentum and for consistency one should thus neglect also those contributions which arise in the effective theory approach by coupling H_d to the sparticle loop. In the diagrammatic calculation this amounts in dropping terms which contain the vev v_d from the squark-, chargino- and neutralino mixing matrices in appendix A.1. Furthermore, in (4.14) we discarded some numerically small contributions stemming from the bino-component of the neutralinos or involving wino-higgsino mixing. The former are suppressed by g'^2/g^2, the latter by $g^2(v^2/M_{\rm SUSY}^2)$.

Whereas in the effective-theory approach the $\tan\beta$-enhancement was easily recognisable by the coupling to H_u, in the diagrammatic treatment it is hidden in the elements of the mixing matrices. Using the analytic expressions for these matrices listed in Appendix A.1, i.e. identities like Eq. (A.7) and Eq. (A.13), we can derive formulae for the gluino- and chargino-contributions in which the $\tan\beta$-enhancement becomes explicit. Writing

$$\Delta_b^K = \epsilon_b^K \tan\beta \qquad \text{for } K = \tilde{g}, \tilde{\chi}^\pm, \tilde{\chi}^0 \qquad \text{and} \qquad \epsilon_b = \epsilon_b^{\tilde{g}} + \epsilon_b^{\tilde{\chi}^\pm} + \epsilon_b^{\tilde{\chi}^0} \tag{4.15}$$

we find

$$\begin{aligned} \epsilon_b^{\tilde{g}} &= -\frac{2\alpha_s}{3\pi} m_{\tilde{g}} \mu^* C_0(m_{\tilde{g}}, m_{\tilde{b}_1}, m_{\tilde{b}_2}), \\ \epsilon_b^{\tilde{\chi}^\pm} &= -\frac{y_t^2}{16\pi^2} A_t^* \mu^* \left(D_2 - |M_2|^2 D_0\right) + \frac{g^2}{16\pi^2} \mu^* M_2^* \left(D_2 - m_{\tilde{t}_R}^2 D_0\right), \end{aligned} \tag{4.16}$$

where $D_{0,2} = D_{0,2}(m_{\tilde{\chi}_1^\pm}, m_{\tilde{\chi}_2^\pm}, m_{\tilde{t}_1}, m_{\tilde{t}_2})$. Eqs. (4.12)-(4.16) generalise the well-known expressions of Ref. [65] to the case of complex MSSM parameters.

4.3.2. Renormalisation of the Yukawa coupling

In the procedure of renormalisation, the fundamental (a priori free) parameters of the Lagrangian are expressed by measured (or at least measurable) quantities. This is done by calculating corresponding observables in terms of the parameters in the Lagrangian, expressing then the Lagrangian in terms of these observables. The form of the final Lagrangian obviously depends on the chosen observables (renormalisation scheme), predictions obtained from it for further observables, however, are scheme independent[3]. To keep things as simple as possible one chooses observables which are closely related to fundamental parameters in the Lagrangian using then these parameters directly as "pseudo-observables".

The Yukawa coupling y_b could most directly be assessed using A^0, H^0 decays into b quarks as observables. The corresponding amplitudes do not receive any $\tan\beta$-enhanced corrections to be subtracted by δy_b and so no $\tan\beta$-enhanced counterterm effects would arise. Having fixed y_b in this way from experiment, we could then predict other observables, among them the b quark mass m_b. Since its tree-level amplitude $m_b = v y_b \cos\beta$ is $\cos\beta$-suppressed, we encounter the type of $\tan\beta$-enhancement discussed in the second paragraph of Section 4.2.1 leading to the enhanced one-loop self-energy contributions (4.11)-(4.16). However, there are no enhancement effects beyond one-loop and so no resummation is needed.

Yet, in any phenomenological application we have to face the fact that we have precise data on m_b but not on the Higgs couplings. Therefore we will rather use the measured b quark mass as observable to renormalise $y_b = m_b/v_d$ to the $\overline{\text{MS}}$ mass m_b. To this end we must calculate SUSY self-energy corrections to m_b in terms of y_b and subtract them by δy_b according to

$$v_d \delta y_b = \delta m_b = -\frac{m_b}{2}\left[\Sigma_b^{LL}(m_b^2) + \Sigma_b^{RR}(m_b^2)\right] - \Sigma_b^{RL}(m_b^2). \tag{4.17}$$

[3] in perturbative calculations the scheme independence holds only up to the considered accuracy.

The self-energy $\Sigma_b^{RL}(m_{\tilde{b}}^2)$ contains $\tan\beta$-enhanced parts and because of our renormalisation condition this $\tan\beta$-enhancement creeps in the Yukawa counterterm δy_b.

The self-energy Σ_b^{RL} is a function of the Yukawa coupling y_b, which comes in either directly via the quark-squark-higgsino-vertex or indirectly via the sbottom mixing angle. We make this y_b-dependence explicit in what follows by writing $\Sigma_b^{RL}(y_b)$. Now, let us consider such self-energy diagrams in which one or more of the couplings y_b are replaced by the counterterm δy_b. Despite being formally of a higher loop order, these diagrams are comparable in size with the original ones because δy_b is of the same order of magnitude as y_b thanks to its $\tan\beta$-enhancement. In order to cancel also the diagrams with counterterm insertion, the Yukawa counterterm δy_b must fulfill

$$v_d \delta y_b = -\Sigma_b^{RL}(y_b + \delta y_b). \tag{4.18}$$

to all orders in the perturbative expansion and to leading order in $\tan\beta$.

Let us adopt the notation from Eqs. (4.11) and (4.15) by writing

$$\Sigma_b^{RL}(y_b + \delta y_b) \;=\; v_d(y_b + \delta y_b)\epsilon_b \tan\beta. \tag{4.19}$$

Whenever ϵ_b does not depend on $y_b^{(0)} \equiv y_b + \delta y_b$, as it is the case for example for $\epsilon_b^{\tilde{g}}$ and $\epsilon_b^{\tilde{\chi}^\pm}$ in Eq. (4.16), one can easily solve (4.18) for δy_b. The resulting resummation formula

$$y_b^{(0)} \;=\; y_b + \delta y_b \;=\; \frac{m_b}{v_d} - \frac{m_b}{v_d}\frac{\epsilon_b \tan\beta}{1+\epsilon_b\tan\beta} \;=\; \frac{m_b}{v_d(1+\epsilon_b\tan\beta)} \tag{4.20}$$

is identical to Eq. (4.5) with the replacement (4.6). Note that in our diagrammatic derivation we made use only of the hierarchy $m_b \ll M_{\text{SUSY}}$ but did not assume any hierarchy between $M_{\text{SUSY}}, M_{A^0,H^0,H^\pm}$ and v.

The solution of Eq. (4.18) automatically resums contributions of the form $(\epsilon_b \tan\beta)^k$ to all orders $k = 1, 2, \ldots$. This resummation can also be performed explicitly by expanding Eq. (4.18) with respect to the loop order and then solving it order by order. Denoting the k-th order counterterm by $\delta y_b^{(k)}$ and assuming again an y_b-independent ϵ_b, we find

$$\begin{aligned} \delta y_b^{(1)} &= -y_b\epsilon_b\tan\beta, \\ \delta y_b^{(k)} &= -\delta y_b^{(k-1)}\epsilon_b\tan\beta \;=\; y_b(-\epsilon_b\tan\beta)^k, \qquad (k = 2, 3, \ldots). \end{aligned} \tag{4.21}$$

This perturbative expansion can immediately be interpreted in terms of Feynman diagrams: The k-th order counterterm $\delta y_b^{(k)}$ has to cancel the insertion of $\delta y_b^{(k-1)}$ into the one-loop self-energy diagram. These are the only possible enhanced higher-order diagrams because of the absence of genuine $\tan\beta$-enhanced multi-loop contributions. The recursive determination of the $\delta y_b^{(k)}$

permits eventually an explicit resummation

$$\delta y_b = \sum_{k=1}^{\infty} \delta y_b^{(k)} = \sum_{k=1}^{\infty} y_b(-\epsilon_b \tan\beta)^k = -\frac{m_b}{v_d}\frac{\epsilon_b \tan\beta}{1+\epsilon_b \tan\beta}. \tag{4.22}$$

yielding the same resummation formula (4.20) for $y_b^{(0)}$ as derived aboved.

4.3.3. Scheme dependence of the resummation formula

It should be stressed that the resulting resummation formula for $y_b^{(0)}$ depends on the renormalisation scheme for the input parameters in the sbottom sector. This is because the y_b-dependence of Σ_b^{RL}, and hence the algebraic equation (4.18) for δy_b, is modified if we change our parameter set. As the masses and mixing angles of SUSY particles have not been measured yet, it is most prominent to take parameters as input which directly appear in the Lagrangian. For the sbottom sector this means choosing $m_{\tilde{b}_{L,R}}$, μ and $\tan\beta$ as input which parametrise the matrix elements of the sbottom mass matrix $M_{\tilde{b}}^2$ displayed in (A.1). With $M_{\tilde{b}}^2$ as input, the quantities $m_{\tilde{b}_{1,2}}$, $\tilde{\theta}_b$ and $\tilde{\phi}_b$ defined in (A.4) and (A.5) are fixed by the diagonalisation procedure, i.e. they are not free parameters but functions of the elements of $M_{\tilde{b}}^2$. On the other hand, assuming that some day it will be possible to measure $m_{\tilde{b}_{1,2}}$, $\tilde{\theta}_b$ and $\tilde{\phi}_b$, one can also take these quantities (or other combinations of parameters) directly as input, i.e. renormalise the Lagrangian to corresponding observables. Note that it is not possible to distinguish between different schemes for the input parameters in the limit $v/M_{\text{SUSY}} \to 0$ because from (A.6), (A.7) we infer

$$m_{\tilde{b}_{1,2}}^2 = m_{\tilde{b}_{L,R}}^2 \left(1+\mathcal{O}\left(v^2/M_{\text{SUSY}}^2\right)\right), \qquad \sin 2\tilde{\theta}_b = \mathcal{O}\left(v/M_{\text{SUSY}}\right). \tag{4.23}$$

Beyond the decoupling limit, however, different choices for the set of input parameters lead to different resummation formulae for $y_b^{(0)}$, as shown in table 4.1 for the numerically dominating gluino part $\Delta_b^{\tilde{g}}$ of Δ_b. In the following, we will deduce these results and discuss also the inclusion of the chargino- and the neutralino-contribution $\Delta_b^{\tilde{\chi}^\pm}$ and $\Delta_b^{\tilde{\chi}^0}$:

(i) **Input:** $m_{\tilde{b}_1}^2, m_{\tilde{b}_2}^2; \mu, \tan\beta$
If we express the sbottom mixing angle $\tilde{\theta}_b$ and phase $\tilde{\phi}_b$ in (4.12) through our input parameters, using relation (A.7), the bottom mass in $\Delta_b^{\tilde{g}}$ in (4.12) cancels and we find the gluino contribution to Σ_b^{RL} to be linear in y_b. This case was studied in the previous section to illustrate the resummation procedure resulting in formula (4.20). Therefore we arrive at

$$y_b^{(0)} = \frac{m_b}{v_d(1+\Delta_b^{\tilde{g}})}. \tag{4.24}$$

If we assume the chargino and neutralino contributions to Σ_b^{RL} to be linear in y_b, too, they

input	resummation formula
$m_{\tilde{b}_1}, m_{\tilde{b}_2}, \tilde{\theta}_b, \tilde{\phi}_b$	$y_b^{(0)} = \frac{m_b}{v_d}\left(1 - \Delta_b^{\tilde{g}}\right)$
$m_{\tilde{b}_1}, m_{\tilde{b}_2}, \mu, \tan\beta$	$y_b^{(0)} = \frac{m_b}{v_d(1+\Delta_b^{\tilde{g}})}$
$m_{\tilde{b}_L}, m_{\tilde{b}_R}, \mu, \tan\beta$	analytic resummation impossible, use iteration.

Table 4.1: resummation formulae for $y_b^{(0)}$ for different choices of the input parameters. Only the gluino contribution $\Delta_b^{\tilde{g}}$ is considered.

can easily be included into (4.24) by the replacement

$$\Delta_b^{\tilde{g}} \to \Delta_b = \Delta_b^{\tilde{g}} + \Delta_b^{\tilde{\chi}^\pm} + \Delta_b^{\tilde{\chi}^0}. \tag{4.25}$$

To this end, $\Delta_b^{\tilde{\chi}^\pm}$ and $\Delta_b^{\tilde{\chi}^0}$ must not depend on y_b. The chargino contribution $\Delta_b^{\tilde{\chi}^\pm}$ in (4.13) is indeed independent of y_b. The neutralino contribution $\Delta_b^{\tilde{\chi}^0}$ in (4.14) can be rewritten as

$$\begin{aligned}\Delta_b^{\tilde{\chi}^0} &= \frac{g^2}{16\pi^2}\frac{1}{\cos\beta}\sum_{m=1}^{4}\frac{m_{\tilde{\chi}_m^0}}{2M_W}\widetilde{N}_{m2}^*\widetilde{N}_{m3}^* \cdot I_2^{(0)}(m_{\tilde{\chi}_m^0}, m_{\tilde{b}_1}) \\ &-\frac{g^2}{16\pi^2}\frac{1}{\cos\beta}\sum_{m=1}^{4}\frac{m_{\tilde{\chi}_m^0}}{2M_W}\widetilde{N}_{m2}^*\widetilde{N}_{m3}^*\sin^2\tilde{\theta}_b\left(I_2^{(0)}(m_{\tilde{\chi}_m^0}, m_{\tilde{b}_1}) - I_2^{(0)}(m_{\tilde{\chi}_m^0}, m_{\tilde{b}_2})\right),\end{aligned} \tag{4.26}$$

where the first line is independent of y_b, but the second line is found to contain terms of second order and higher in y_b after insertion of (A.7). In the decoupling limit $M_{\text{SUSY}} \gg v$, these higher-order terms, which are proportional to $\sin^2\tilde{\theta}_b \propto v^2/M_{\text{SUSY}}^2$, vanish, and the neutralino contribution is correctly included by the replacement rule (4.25). For $M_{\text{SUSY}} \sim v$ on the other hand, the higher order terms spoil the proper resummation because equation (4.18) cannot be solved analytically anymore. As $\Delta_b^{\tilde{\chi}^0}$ is small anyway and the second line of (4.26) suffers in addition from a GIM-like suppression, formula (4.24) with the replacement (4.25), though not entirely correct in this case, still holds to a very good approximation.

(ii) **Input:** $m_{\tilde{b}_1}^2, m_{\tilde{b}_2}^2; \tilde{\theta}_b, \tilde{\phi}_b$
Assuming that some day it will be possible to measure $\tilde{\theta}_b$ and $\tilde{\phi}_b$, we could take these quantities as our input instead of μ and $\tan\beta$. In (4.12) $\Delta_b^{\tilde{g}}$ is directly given as a function

of $\tilde{\theta}_b$ and $\tilde{\phi}_b$. Obviously, $\Sigma_b^{RL,\tilde{g}} = y_b v_d \Delta_b^{\tilde{g}}$ does not exhibit any explicit y_b-dependence in this case, so that no reinsertion of δy_b into $\Sigma_b^{RL,\tilde{g}}$ is possible. The counterterm δy_b is thus an ordinary one-loop counterterm and the modified relation between $y_b^{(0)}$ and m_b reads

$$y_b^{(0)} = \frac{m_b}{v_d}(1 - \Delta_b^{\tilde{g}}). \tag{4.27}$$

On the other hand, the chargino and neutralino contributions $\Sigma_b^{RL,\tilde{\chi}^\pm} = y_b v_d \Delta_b^{\tilde{\chi}^\pm}$ and $\Sigma_b^{RL,\tilde{\chi}^0} = y_b v_d \Delta_b^{\tilde{\chi}^0}$ in (4.13) and (4.14) seem to be linear in y_b suggesting therefore their inclusion via

$$y_b^{(0)} = \frac{m_b}{v_d} \frac{1 - \Delta_b^{\tilde{g}}}{1 + \Delta_b^{\tilde{\chi}^\pm} + \Delta_b^{\tilde{\chi}^0}}. \tag{4.28}$$

However, there is a subtlety: In the large-$\tan\beta$ limit, Eq. (A.7) implies a correlation between y_b and μ:

$$e^{i\tilde{\phi}_b} \sin 2\tilde{\theta}_b = -\frac{2 y_b^{(0)*} \mu v_u}{m_{\tilde{b}_1}^2 - m_{\tilde{b}_2}^2}. \tag{4.29}$$

This equation implicitly defines μ in scheme (ii) and it follows that μ then inherits the large corrections from $y_b^{(0)}$. Since μ enters the chargino and neutralino mass matrices $M_{\tilde{\chi}^{\pm,0}}$ in Eqs. (A.9) and (A.14), expressing it in terms of $y_b^{(0)}$ via Eq. (4.29) leads to a very involved y_b-dependence of $\Delta_b^{\tilde{\chi}^\pm}$ and $\Delta_b^{\tilde{\chi}^0}$ encoded in the masses $m_{\tilde{\chi}^\pm_{1,2}}$, $m_{\tilde{\chi}^0_{1..4}}$ and mixing matrices $\widetilde{U}$, $\widetilde{V}$ and $\widetilde{N}$. This situation renders an analytic resummation of the chargino and neutralino contributions impossible and calls for the following iterative procedure: One calculates $y_b^{(0)}$ from (4.28) in terms of an initial μ value, determines then an updated value for μ from (4.29) and repeats these steps until Eqs. (4.28) and (4.29) are sufficiently compatible.

(iii) **Input:** $m_{\tilde{b}_L}^2$, $m_{\tilde{b}_R}^2$; μ, $\tan\beta$
As the masses and mixing angles of the SUSY particles are not measured yet, this set is the most prominent one because its elements directly appear in the Lagrangian. In terms of these input parameters, the mixing angle can be expressed with help of the relation

$$\tan 2\tilde{\theta}_b = -\frac{y_b v_u \mu}{(m_{\tilde{b}_L}^2 - m_{\tilde{b}_R}^2)}. \tag{4.30}$$

Since $\Delta_b^{\tilde{g}}$ is proportional to $\sin 2\tilde{\theta}_b = \tan 2\tilde{\theta}_b / (\sqrt{1 + \tan^2 2\tilde{\theta}_b})$, and in addition the squark masses appearing in the loop functions have to be replaced by $m_{\tilde{b}_L}^2$ and $m_{\tilde{b}_R}^2$ via (A.6), the y_b-dependence of $\Delta_b^{\tilde{g}}$ gets so complicated that (4.18) cannot be solved analytically anymore. This problem can be avoided in the following way: In a first approximation, we determine $m_{\tilde{b}_{1,2}}^2$ from (A.6) using the tree level value for y_b. Now we can calculate Δ_b as a function of the parameter set (i). In a next step, the resulting modified Yukawa coupling (4.24) can be reinserted into (A.6) to get corrected values for $m_{\tilde{b}_{1,2}}^2$. This procedure has

Figure 4.4: Self-energy and counterterm insertion into internal b quark line

to be repeated until the value of Δ_b converges. The resummed Yukawa coupling is then given by (4.24). Alternatively, we could calculate $\Delta_b^{\tilde{g}}$ and $\Delta_b^{\tilde{\chi}^0}$ iteratively as a function of the input parameters (ii), determining $\sin 2\tilde{\theta}_b$ from (4.30). In that case, equation (4.28) provides the resummed Yukawa coupling.

Eq. (4.24) has the same form as the widely-used relation between $y_b^{(0)}$ and m_b valid in the decoupling limit and quoted in (4.5). Therefore we will take parameter set (i) as the physical input from now on.

4.3.4. Self-energies in internal quark lines

Before passing on to the flavour-changing case, let us briefly discuss the situation of self-energy subdiagrams in internal quark lines of Feynman diagrams (left diagram in Fig. 4.4). Such diagrams are the only potential source of flavour-conserving $\tan\beta$-enhanced loop corrections to a Feynman amplitude $\mathcal{M}$ since, as we argued in Section 4.2.1, genuine diagrams cannot be enhanced. It is important to notice that the self-energy correction is only $\tan\beta$-enhanced with respect to the single propagator diagram if the additional propagator generates an inverse power of $m_b = y_b v \cos\beta$. Such a $1/m_b$-behaviour can only arise if the momentum p flowing through the propagator is of the order m_b or lower implying $p^2 \ll M_{\rm SUSY}^2$. However, we have constructed the mass counterterm $\delta m_b = v_d \delta y_b$ in Section 4.3.2 in such a way that its insertion (right diagram in Fig. 4.4) subtracts the self-energy correction in this momentum region.

The only exceptional case in which self-energy insertions into internal quark lines become relevant are processes for which higher orders in the ratio $m_b/M_{\rm SUSY}$ have to be considered. In this situation the right-hand side in (4.17) has to be expanded to higher orders in $m_b/M_{\rm SUSY}$ in order to find the appropriate counterterm δm_b, whereas only the leading term was kept in Section 4.3.2. We stress that this expansion does not spoil the resummation of the counterterm. This means δm_b is obtained by calculating ϵ_b to the desired order in $m_b/M_{\rm SUSY}$ and expanding (4.22) in this ratio. For example writing $\epsilon_b = \epsilon_b^{(0)} + \epsilon_b^{(2)}$ with $\epsilon_b^{(0)} = \mathcal{O}(m_b^0/M_{\rm SUSY}^0)$ and $\epsilon_b^{(2)} = \mathcal{O}(m_b^2/M_{\rm SUSY}^2)$ we find

$$\delta m_b = -m_b \frac{\epsilon_b^{(0)} \tan\beta}{1+\epsilon_b^{(0)}\tan\beta} - m_b \frac{\epsilon_b^{(2)}\tan\beta}{\left(1+\epsilon_b^{(0)}\tan\beta\right)^2} . \qquad (4.31)$$

Now let us assume that the internal quark propagator which is subject to the self-energy and counterterm insertions in Fig. 4.4 carries momentum $p^2 = 0$. Then the self-energy and counterterm insertions only partially cancel because δm_b is determined at $p^2 = m_b^2$ while the self-energy is probed at $p^2 = 0$. Therefore the second term in (4.31) survives and its insertion generates a $\tan\beta$-enhanced correction of higher order in m_b/M_{SUSY}. An important physical process for which this situation occurs is $b \to s\gamma$ where the leading order is $\mathcal{O}(m_b^2/M_{\text{SUSY}}^2)$. It will be discussed in Section 4.4.2 in the context of flavour-changing enhanced corrections.

4.4. Flavour mixing I: External leg corrections

In the previous sections we discussed the consequences of $\tan\beta$-enhanced contributions to the b-quark self-energy Σ_b^{RL}. Let us now have a look at the corresponding flavour-changing self-energies Σ_{ij}^{RL}. We will see that insertions of Σ_{ij}^{RL}-subdiagrams into external quark legs lead to enhanced corrections to matrix elements $\mathcal{M}$ of FCNC processes and induce enhanced counterterms δV_{ij} to CKM elements.

4.4.1. Flavour-changing $\tan\beta$-enhanced self-energies

In the framework of naive MFV the gluino and neutralino couplings to (s)quarks are flavour-diagonal at tree-level. Therefore only chargino diagrams generate $\tan\beta$-enhanced contributions to Σ_{ij}^{RL} (see Fig. 4.5). In the case of d-s transitions, the stop contribution is suppressed by the tiny CKM combination $V_{ts}^* V_{td}$. If we neglect in addition the small up and charm Yukawa couplings, the chargino couples only to left-handed squarks of the first two generations. Furthermore, switching off these Yukawa couplings in our naive MFV scenario restores a $SU(2)$ symmetry for the left-handed $\tilde{u}$- and $\tilde{c}$-squarks causing a GIM cancellation of their contributions to d-s transitions. Therefore with these approximations only self-energies involving a bottom quark survive and MFV dictates their form:

$$\bar{d}_R^i \, (Y_d Y_u^\dagger Y_u)_{ij} \, Q_j \qquad \longrightarrow \qquad \Sigma_{ij}^{RL} \sim y_{d_i} \, y_t^2 \, V_{ti}^* V_{tj} \, . \tag{4.32}$$

The explicit expression for Σ_{ij}^{RL} reads

$$\Sigma_{ij}^{RL} = V_{ti}^* V_{tj} \frac{m_i \Delta_{\text{FC}}}{1+\Delta_i}, \qquad \text{for} \quad (i,j) = (3,1), (3,2), (1,3), (2,3). \tag{4.33}$$

with

$$\Delta_{FC} = \Delta_b^{\tilde{\chi}^\pm} + \Delta_0^{\tilde{\chi}^\pm} \tag{4.34}$$

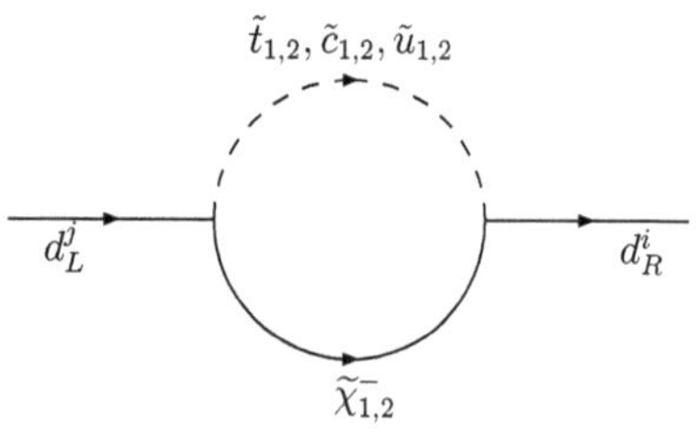

Figure 4.5: $\tan\beta$-enhanced flavour-changing self-energy diagram

and

$$\Delta_0^{\tilde{\chi}^\pm} = -\frac{g^2}{16\pi^2}\frac{1}{\cos\beta}\sum_{m=1}^{2}\frac{m_{\tilde{\chi}_m^\pm}}{\sqrt{2}M_W}\,\widetilde{U}_{m2}^*\widetilde{V}_{m1}^*\,B_0(m_{\tilde{\chi}_m^\pm}, m_{\tilde{q}})\,. \tag{4.35}$$

Here $m_{\tilde{q}}$ denotes the common mass of left-handed up- and charm-squarks. The expression for $\Delta_b^{\tilde{\chi}^\pm}$ has already been given in (4.13). The factor $1/(1+\Delta_i)$ in (4.33) accounts for the renormalisation of the Yukawa coupling y_{d_i} such that m_i is the $\overline{\text{MS}}$-mass of the quark d_i. Its appearance is mandatory in MFV as we have seen in (4.32) and it causes self-energies with right-handed d- or s-quarks to be suppressed by m_i/m_b ($i = d, s$) compared to their chirality-exchanged counterparts. As in the flavour-conserving case, we can make use of Eqs. (A.7) and (A.13) to rewrite Δ_{FC} as

$$\Delta_{\text{FC}} = \epsilon_{\text{FC}}\tan\beta \tag{4.36}$$

with

$$\epsilon_{\text{FC}} = -\frac{y_t^2}{16\pi^2}A_t^*\mu^*\left(D_2 - |M_2|^2 D_0\right) + \frac{g^2}{16\pi^2}M_2^*\mu^*\left(D_2 - m_{\tilde{t}_R}^2 D_0 - C_0\right) \tag{4.37}$$

where $D_{0,2} = D_{0,2}(m_{\tilde{\chi}_1^\pm}, m_{\tilde{\chi}_2^\pm}, m_{\tilde{t}_1}, m_{\tilde{t}_2})$ and $C_0 = C_0(m_{\tilde{\chi}_1^\pm}, m_{\tilde{\chi}_2^\pm}, m_{\tilde{q}})$. Whereas the first term in (4.37) displays an explicit factor of y_t^2 and thus apparently obeys the MFV structure of (4.32), this is less obvious for the second term. Note, however, that this part of ϵ_{FC} vanishes due to the universality of the left-handed squark mass terms in naive MFV if the left-right mixing of the top squarks is neglected. Therefore it is proportional to $y_t^2\, v^2/M_{\text{SUSY}}$ and this implies its absence in the decoupling limit[4].

4.4.2. Flavour-changing self-energies in external quark legs

Let us consider the generic situation of a flavour-changing self-energy subdiagram in an external quark leg of some Feynman diagram, as displayed in Fig. 4.6. For definiteness we focus on b-s

[4] In the more general symmetry-based definition of MFV the left-handed stop mass term may differ from those of the first two generations by a term proportional to y_t^2 due to the b_1-term in (3.12). In such a scenario the g^2-part in (4.37) would be present also in the decoupling limit.

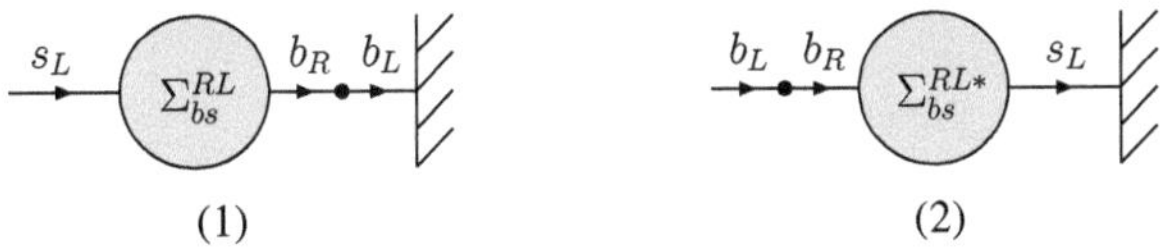

Figure 4.6: Feynman diagrams with flavour-changing self-energy in an external leg.

transitions, corresponding results for b-d transitions are obtained by obvious replacements. Unlike insertions of flavour-conserving self-energies into external legs which have to be truncated[5], insertions of flavour-changing self-energies can be treated as one-particle irreducible (1PI) [66]. The only prerequisite for this treatment is that the mass difference $m_b - m_s$ is much larger than the self-energy $|\Sigma_{bs}^{RL}|$ and it is certainly fulfilled because of the CKM-suppression of Σ_{bs}^{RL} (even though the loop-suppression might be lifted by a large factor $\tan\beta$).

Setting $m_s = 0$ we find that the amplitudes of the Feynman diagrams in Fig. 4.6 are given by

$$\mathcal{M}_1 = \mathcal{M}_1^{\text{rest}} \cdot \left.\frac{i(\not{p} + m_b)}{p^2 - m_b^2}\right|_{\not{p}=0} (-i\Sigma_{bs}^{RL}) = -\mathcal{M}_1^{\text{rest}} \cdot V_{ts} V_{tb}^* \frac{\epsilon_{\text{FC}} \tan\beta}{1 + \epsilon_b \tan\beta}, \tag{4.38}$$

$$\mathcal{M}_2 = \mathcal{M}_2^{\text{rest}} \cdot \left.\frac{i(\not{p} + m_s)}{p^2 - m_s^2}\right|_{\not{p}=m_b^{\text{pole}}} (-i\Sigma_{bs}^{RL*}) = +\mathcal{M}_2^{\text{rest}} \cdot V_{ts}^* V_{tb} \frac{\epsilon_{\text{FC}}^* \tan\beta}{1 + \epsilon_b^* \tan\beta}, \tag{4.39}$$

where $\mathcal{M}_i^{\text{rest}}$ stands for the part of the Feynman amplitude corresponding to the truncated diagram. The expressions (4.38) and (4.39) are of order $\mathcal{O}(\epsilon_{\text{FC}} \tan\beta)$. Thus, if $\tan\beta$ is large enough to compensate for the loop-factor ϵ_{FC}, it is possible to get a $b \to s$ transition without paying the price of a loop suppression. In this way one finds $\tan\beta$-enhanced corrections to FCNC $b \to s$ processes by sourcing the flavour change out into an external quark leg rendering the underlying loop-diagram flavour-diagonal. As an example consider the SUSY-contribution to $b \to s\gamma$: For large $\tan\beta$ the one-loop amplitude is dominated by the chargino-stop diagram giving a contribution of the form (loop$\times \tan\beta$). Taking into account the above mentioned external leg corrections one finds contributions involving gluino-sbottom loops, like the diagrams in Fig. 4.7, which are of the form (loop $\times$ $\tan\beta)^2$.

Moreover, we encounter in $b \to s\gamma$ the situation discussed in Section 4.3.4: Expanding the diagrams in m_b/M_{SUSY}, the first non-vanishing contribution is found at second order in this ratio. To this order the two diagrams on the left in Fig. 4.7 do not cancel completely (see discussion in Section 4.3.4) and the remnant complements the vertex diagram on the upper right to a gauge-invariant result[6].

[5]they instead enter the S-matrix elements through the LSZ factor.

[6]For completeness, also the insertion of a wave function counterterm is shown in Fig. 4.7 (lower right diagram). It is needed to cancel some non-$\tan\beta$-enhanced contributions.

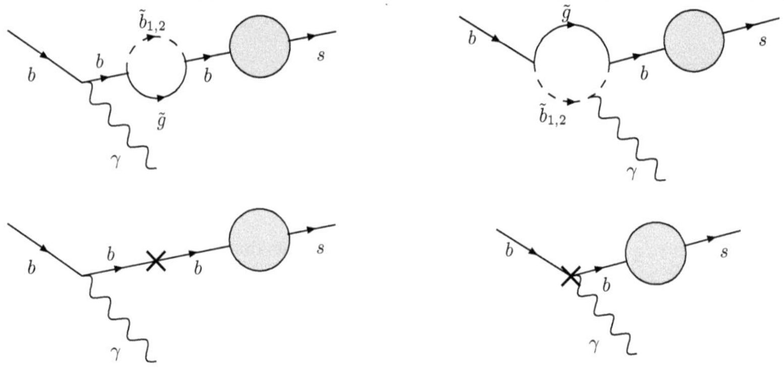

Figure 4.7: Some diagrams with self-energies in external lines for the process $b \to s\gamma$

In this context it is natural to ask whether the generation of $\tan\beta$-enhanced $b \to s$ transitions via self-energy insertions also occurs for internal quark lines. In other words, we interested in the flavour-changing counterpart of the diagram on the left in Fig. 4.4. In principle it is possible to construct such diagrams and, furthermore, there is no counterterm to cancel them, in contrast to the flavour-conserving case. However, as in the flavour-conserving case, the momentum flowing through the propagator containing the self-energy subdiagram has to be of order m_b or smaller. Since we are not aware of a meaningful physical process developing this situation, we do not consider this possibility further.

4.4.3. QCD corrections to flavour-changing self-energies

Before investigating the further consequences of the $\tan\beta$-enhanced flavour transitions, we want to point out a subtlety of Eq. (4.39). The b-quark mass which enters the propagator via the equation of motion is the pole mass m_b^{pole}. The b-quark mass appearing in Σ_{bs}^{RL}, on the other hand, is the $\overline{\text{MS}}$-mass m_b. However, if QCD-corrections to the diagrams of Fig. 4.6 are taken into account, additional contributions add to the $\overline{\text{MS}}$-mass in Σ_{bs}^{RL} to give the pole mass m_b^{pole}. Therefore the b-quark mass correctly cancels in (4.39).

To see this we consider an effective theory at $\mu \sim \mathcal{O}(m_b)$ where the SUSY-particles are integrated out. The self-energy Σ_{bs}^{RL} then appears as Wilson coefficient of the (on-shell vanishing) operator $\bar{b}P_L s$. Comparing QCD corrections to this operator to QCD corrections to the bottom mass m_b (see Fig. 4.8) we find

$$\frac{\Sigma_{bs}^{RL(1)}(p)}{\Sigma_{bs}^{RL}} = \frac{\Sigma_b^{QCD}(p)}{m_b}, \tag{4.40}$$

where p denotes the external momentum. Therefore the Wilson coefficient Σ_{bs}^{RL} and the $\overline{\text{MS}}$-mass

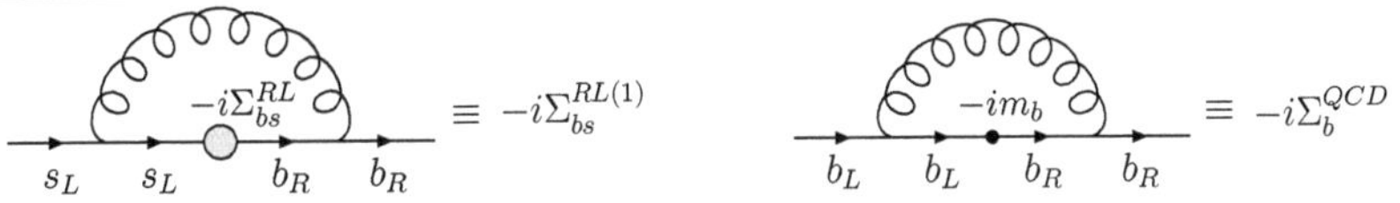

Figure 4.8: QCD corrections to the self-energy Σ_{bs}^{RL} (left) and the bottom mass m_b (right).

m_b renormalise the same way. To make the behaviour under renormalisation explicit we write

$$\Sigma_{bs}^{RL} = m_b A \tag{4.41}$$

where now A is renormalisation-scale-independent (note the analogy to the definitions of Δ_b and Δ_{FC} in Eqs. (4.11) and (4.33) which are thus renormalisation-scale independent).

Now we calculate QCD corrections to the diagrams in Fig. 4.6. Using the parametrisation (4.41) for Σ_{bs}^{RL} and neglecting the s-quark mass, the Feynman amplitudes for the diagrams in Fig. 4.6 read

$$\mathcal{M}_1^{(1)} = \mathcal{M}_1^{\text{rest}} \cdot \left.\frac{i(\not{p}+m_b)}{p^2-m_b^2}\right|_{\not{p}=0} (-i\Sigma_{bs}^{RL}) = -\mathcal{M}_1^{\text{rest}} \cdot A, \tag{4.42}$$

$$\mathcal{M}_2^{(2)} = \mathcal{M}_2^{\text{rest}} \cdot \left.\frac{i(\not{p}+m_s)}{p^2-m_s^2}\right|_{\not{p}=m_b^{\text{pole}}} (-i\Sigma_{bs}^{RL*}) = +\mathcal{M}_2^{\text{rest}} \cdot A^* \frac{m_b}{m_b^{\text{pole}}}. \tag{4.43}$$

Since we want to perform a calculation up to order α_s in the effective theory we have to determine A from two-loop matching at the SUSY scale and we make this explicit by writing

$$A = A^{(0)} + A^{(1)} \tag{4.44}$$

where $A^{(1)}$ contains $\mathcal{O}(\alpha_s)$ QCD-corrections. The one-loop corrections to $\mathcal{M}_1$ and $\mathcal{M}_2$ in the effective theory are given in Figs. 4.9 and 4.10, respectively, with diagrams (1b) and (2b) taking into account the counterterm to the Wilson coefficient $\Sigma_{bs}^{RL} = m_b\, A$. As a consequence of (4.40), the contributions of (1a) and (1c) and of (1b) and (1d) cancel pairwise. Therefore the expression for $\mathcal{M}_1$ in (4.42) still holds at one loop, yet with $A = A^{(0)} + A^{(1)}$ instead of $A = A^{(0)}$. For the contributions of (2a) and (2b) we find with help of (4.40)

$$\mathcal{M}_2^{(2a)} = \mathcal{M}_2^{\text{rest}} \cdot \frac{i(\not{p}+m_s)}{p^2-m_s^2} \left.\left(-i\Sigma_{bs}^{RL(1)*}(p)\right)\right|_{\not{p}=m_b^{\text{pole}}} = \mathcal{M}_2^{\text{rest}} \cdot A^{(0)*} \left.\frac{\Sigma_b^{QCD}(p)}{m_b^{\text{pole}}}\right|_{\not{p}=m_b^{\text{pole}}} \tag{4.45}$$

$$\mathcal{M}_2^{(2b)} = \mathcal{M}_2^{\text{rest}} \cdot \left.\frac{i(\not{p}+m_s)}{p^2-m_s^2}\right|_{\not{p}=m_b^{\text{pole}}} (-i\delta m_b A^{(0)*}) = \mathcal{M}_2^{\text{rest}} \cdot A^{(0)*} \frac{\delta m_b}{m_b^{\text{pole}}}. \tag{4.46}$$

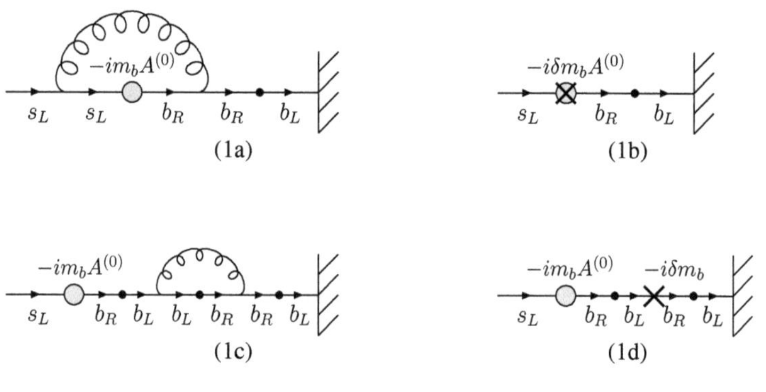

Figure 4.9: QCD corrections to diagram (1) in Fig. 4.6.

Adding these corrections to (4.43) one gets

$$\mathcal{M}_2 = \mathcal{M}_2^{(2)} + \mathcal{M}_2^{(2a)} + \mathcal{M}_2^{(2b)} = \mathcal{M}_2^{\text{rest}} \cdot \frac{A^{(0)*}}{m_b^{\text{pole}}} \left(m_b + m_b \frac{A^{(1)*}}{A^{(0)*}} + \Sigma_b^{QCD}(p)\Big|_{\not{p}=m_b^{\text{pole}}} + \delta m_b \right) . \tag{4.47}$$

Plugging in

$$m_b^{\text{pole}} = m_b + \Sigma_b^{QCD}(p)\Big|_{\not{p}=m_b^{\text{pole}}} + \delta m_b \tag{4.48}$$

and dropping terms of order $\mathcal{O}(\alpha_s^2)$ we get the final result

$$\mathcal{M}_2 = \mathcal{M}_2^{\text{rest}} \cdot (A^{(0)*} + A^{(1)*}) = \mathcal{M}_2^{\text{rest}} \cdot A^* , \tag{4.49}$$

which now does not depend on m_b^{pole} anymore.

Applying this result to our case by expressing A in (4.49) through Σ_{bs}^{RL} via Eqs. (4.41) and (4.33) we find (4.39). Since (4.33) is linear in m_b, the parametrisation of (4.41) is quite natural. When one considers a more general Σ_{bs}^{RL} which is no longer linear in m_b (for example in the generic MSSM), the parameter A depends on m_b via (4.41) but in any case it does not involve m_b^{pole}.

4.4.4. Renormalisation of the CKM matrix

The external leg corrections discussed in the last two sections affect also the u_i-d_j-W^+ coupling if it involves an external d_j-quark (left diagram in Fig. 4.11). Therefore, to be able to extract the CKM matrix elements from a low energy measurement of this coupling, one has to subtract the enhanced corrections by appropriate CKM counterterms δV_{ij}. This amounts to the on-shell

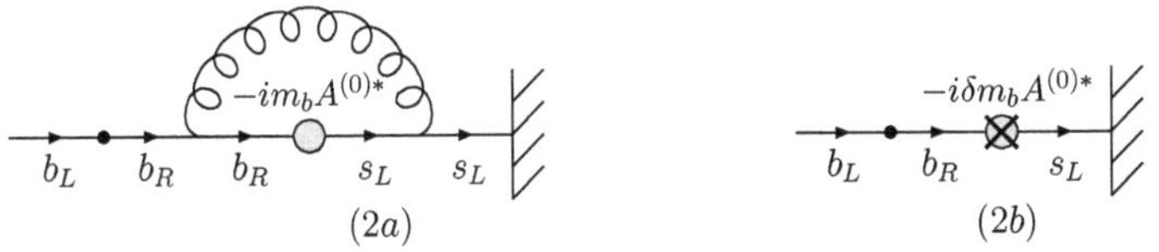

Figure 4.10: QCD corrections to diagram (2) in Fig. 4.6.

renormalisation condition proposed in Ref. [67] and depicted in Fig. 4.11[7]. We find

$$\begin{aligned} \delta V_{ij} &= -V_{ik}\Lambda_{kj}, \qquad \text{with} \\ \Lambda_{kj}(V) &= \begin{cases} \dfrac{m_{d_j}}{m_{d_j}^2 - m_{d_k}^2}\Sigma_{kj}^{LR} + \dfrac{m_{d_k}}{m_{d_j}^2 - m_{d_k}^2}\Sigma_{kj}^{RL} & , k \neq j \\ 0 & , k = j \end{cases} \end{aligned} \tag{4.50}$$

In the following we will neglect contributions to δV_{ij} which are CKM suppressed compared to V_{ij}, i.e. which involve more powers of the Wolfenstein parameter λ than V_{ij}. On the other hand, δV_{ij} never involves less powers of λ than V_{ij} because the MFV framework guarantees that loop corrections cannot lift the CKM-suppression of the u_i-d_j-W^+ vertex.

Note also that our renormalisation prescription preserves the unitarity of the CKM matrix since

$$V^{(0)} = V + \delta V = V(1 - \Lambda) \approx Ve^{-\Lambda}. \tag{4.51}$$

with the anti-hermitian matrix Λ. The last equation in (4.51) holds because each non-vanishing elements of Λ involves at least one power of λ and we neglect subleading powers in λ.

From Eq. (4.33) we find the counterterms δV_{td}, δV_{ts}, δV_{ub} and δV_{cb} to be of order $\mathcal{O}(\epsilon_{\mathrm{FC}} \tan\beta)$ so that they can be comparable in size to the corresponding tree-level quantities V_{ij}. Furthermore, the loop corrections are functions of the CKM matrix elements V_{ij} which enter the three vertices in the left diagram in Fig. 4.11. In Section 4.3.2 we faced an analogous situation in the context of the renormalisation of the Yukawa coupling y_b and all the consequences discussed there apply also here: Reinsertion of the counterterms δV_{ij} into the left diagram of Fig. 4.11 leads to contributions which are formally of higher loop order but also of higher order in $\tan\beta$. To subtract also these higher-order corrections, Eq. (4.50) has to be extended to all orders in the perturbative

[7] In Ref. [68] it has been argued that the on-shell prescription does not lead to gauge-independent results and one should rather use the symmetric subtraction point $\not{p} = 0$. Note that this subtlety does not matter in our case and the two schemes are equivalent since we neglect external quark momenta in the calculation of the SUSY self-energy contributions.

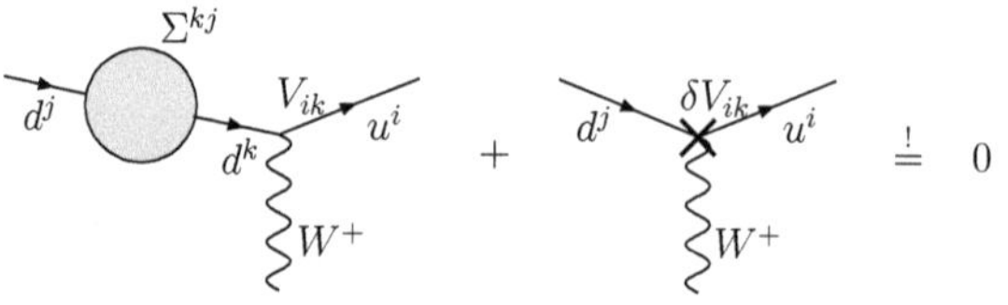

Figure 4.11: On-shell renormalisation of the CKM matrix

expansion. Making the CKM-dependence of the matrix Λ explicit by writing $\Lambda(V)$, we arrive at

$$\delta V_{ij} = -(V_{ik} + \delta V_{ik}) \cdot \Lambda_{kj}(V + \delta V), \tag{4.52}$$

which is in complete analogy to Eq. (4.18) for δy_b. The δV_{ij} are then determined using one of the two methods discussed in Section 4.3.2: Either the matrix equation (4.52) is solved directly for the resummed δV_{ij}, or an order-by-order recursion relation is derived for the $\delta V_{ij}^{(n)}$ followed by an explicit resummation. Of course, both methods lead to the same result which is

$$V^{(0)} = V + \delta V = \begin{pmatrix} V_{ud} & V_{us} & K^* V_{ub} \\ V_{cd} & V_{cs} & K^* V_{cb} \\ K V_{td} & K V_{ts} & V_{tb} \end{pmatrix}, \qquad \text{with} \quad K = \frac{1 + \epsilon_b \tan\beta}{1 + (\epsilon_b - \epsilon_{\text{FC}})\tan\beta}. \tag{4.53}$$

Eq. (4.53) generalises the well-known result from Ref. [12] beyond the decoupling limit $M_{\text{SUSY}} \gg v, M_{A^0,H^0,H^\pm}$ and provides the analytic expression for the result to which the iterative methods proposed in Refs. [23, 24] converge. The resulting resummation formula has the same form as the one valid in the decoupling limit but with ϵ_b and ϵ_{FC} now containing decoupling contributions, too. Furthermore, our resummation formula is valid also in the presence of complex SUSY parameters and our analytic derivation permits an explicit resummation.

4.5. Flavour mixing II: Flavour-changing wave-function counterterms

As we have seen, in order to account for $\tan\beta$-enhanced corrections in flavour-violating processes, one has to consider self-energy insertions into external down-quark legs. Alternatively, since $p^2 \ll M_{\text{SUSY}}^2$ for a momentum p of an on-shell down-type quark and since thus the external leg corrections are local, it is possible to promote their effects to effective flavour-changing vertices, even beyond the decoupling limit because no assumption about a hierarchy between M_{SUSY} and the electroweak scale v is needed. This is illustrated in Fig. 4.12 for the quark-squark-gluino

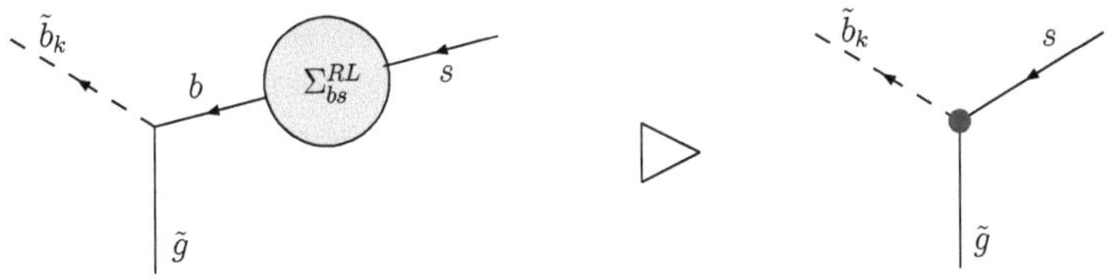

Figure 4.12: FCNC gluino coupling for an on-shell s-quark induced by the $\tan\beta$-enhanced self-energy Σ^{RL}_{bs}

coupling. In addition to the FCNC couplings of neutral Higgs bosons, which are well-studied within the decoupling limit [17], also FCNC couplings of the gluino and neutralino arise in this way. A complete set of Feynman rules for these vertices, including contributions of the form $(\text{loop} \times \tan\beta)^n$ to all orders $n = 1, 2, ...$, is given in Appendix A.3.

Technically, the promotion of the external leg corrections to effective vertices is done by performing a matrix-valued wave-function renormalisation. We discuss the determination and resummation of the corresponding wave-function counterterms in Section 4.5.1 before we turn to the formulation of Feynman rules for the resulting effective couplings in Section 4.5.2.

4.5.1. Flavour-changing wave-function renormalisation

Left- and right-handed up- and down-quark fields are vectors in separate three dimensional family spaces. Their description allows for an arbitrary choice of basis and a change of basis affects the quark fields in form of a matrix-valued wave-function renormalisation. We renormalise the down-quark fields according to

$$d^{(0)}_{i,L} = \left(\delta_{ij} + \frac{1}{2}\delta Z^L_{ij}\right) d_{j,L}, \qquad d^{(0)}_{i,R} = \left(\delta_{ij} + \frac{1}{2}\delta Z^R_{ij}\right) d_{j,R} \tag{4.54}$$

with the anti-hermitian wave-function counterterms

$$\delta Z^L_{ij} = -\delta Z^{L*}_{ji}, \qquad \delta Z^R_{ij} = -\delta Z^{R*}_{ji}. \tag{4.55}$$

Therefore our wave-function renormalisation corresponds to a unitary transformation in flavour space[8]. Note that unlike the physical CKM rotation which connects two sets of physical states, the weak eigenstates and the mass eigenstates, the wave-function renormalisation is unphysical since it only manipulates the coordinate system used for the description of the physical states.

[8]The freedom of normalising the basis vectors corresponds to the ordinary flavour-diagonal wave-function renormalisation which we will not need here.

Any meaningful theory has to be invariant under such a transformation so that no effects on physical observables can arise. This is realised for the S-matrix elements of a quantum field theory by the fact that the wave-function renormalisation drops out from the LSZ formula. In particular, we stress that separate wave-function renormalisation for the quark- and squark-field does not spoil supersymmetry: Whereas the CKM rotation has to be performed on the whole superfields in order not to disjoin the superpartners, an individual wave-function rotation of the quark fields does not rotate them away from their superpartners, it only changes the coordinate system used for their description leaving the physical states unchanged.

In terms of renormalised fields the down-quark mass terms read

$$\mathcal{L}_m = -\, m_{d_j}^{(0)} \bar{d}_{j,R}^{(0)} d_{j,L}^{(0)} + \text{h.c.} = -\left[m_{d_j}^{(0)} \delta_{jk} + \frac{1}{2} m_{d_j}^{(0)} \delta Z_{jk}^{L} - \frac{1}{2} m_{d_k}^{(0)} \delta Z_{jk}^{R} \right] \bar{d}_{j,R} d_{k,L} + \text{h.c.} \quad (4.56)$$

with $m_{d_i}^{(0)} = v_d y_{d_i}^{(0)}$ containing the $\tan\beta$-enhanced corrections determined in Section 4.3. The counterterms $\delta Z_{ij}^{L,R}$ can then be chosen in such a way that their insertion into external quark legs cancels the corresponding self-energy insertion for on-shell momenta of the quark. To this end $\delta Z_{ij}^{L,R}$ must fulfill

$$\Sigma_{ij}^{RL} + m_{d_i}^{(0)} \frac{\delta Z_{ij}^{L}}{2} - m_{d_j}^{(0)} \frac{\delta Z_{ij}^{R}}{2} = 0, \qquad i \neq j\,. \qquad (4.57)$$

Note that for a certain pair (i, j) the two equations obtained for $i \leftrightarrow j$ fix both δZ_{ij}^{L} and δZ_{ij}^{R} since $\delta Z_{ij}^{L,R} = -\delta Z_{ji}^{L,R*}$ but the two self-energies Σ_{ij}^{RL} and Σ_{ji}^{RL} are not related to each other. Solving Eq. (4.57) for $\delta Z_{ij}^{L,R}$ we obtain

$$\frac{\delta Z_{ij}^{L}}{2} = \frac{m_{d_i}^{(0)*} \Sigma_{ij}^{RL} + m_{d_j}^{(0)} \Sigma_{ij}^{LR}}{|m_{d_j}^{(0)}|^2 - |m_{d_i}^{(0)}|^2}, \qquad \frac{\delta Z_{ij}^{R}}{2} = \frac{m_{d_i}^{(0)} \Sigma_{ij}^{LR} + m_{d_j}^{(0)*} \Sigma_{ij}^{RL}}{|m_{d_j}^{(0)}|^2 - |m_{d_i}^{(0)}|^2}, \qquad i \neq j. \qquad (4.58)$$

The anti-hermiticity of the $\delta Z_{ij}^{L,R}$ allows us to restrict the determination of the $\delta Z_{ij}^{L,R}$ to the case $i > j$. Neglecting contributions which are subleading in the small quark mass ratio $m_{d_j}^{(0)}/m_{d_i}^{(0)}$, Eq. (4.58) simplifies to

$$\frac{\delta Z_{ij}^{L}}{2} = -\frac{\Sigma_{ij}^{RL}}{m_{d_i}^{(0)}}, \qquad \frac{\delta Z_{ij}^{R}}{2} = -\frac{\Sigma_{ij}^{LR}}{m_{d_i}^{(0)*}} - \frac{m_{d_j}^{(0)*}}{m_{d_i}^{(0)*}} \frac{\Sigma_{ij}^{RL}}{m_{d_i}^{(0)}}, \qquad i > j. \qquad (4.59)$$

From expression (4.33) for the self-energy Σ_{ij}^{RL} it becomes obvious that both δZ_{ij}^{L} and δZ_{ij}^{R} are $\tan\beta$-enhanced but that δZ_{ij}^{R} is suppressed by one power in $m_{d_j}^{(0)}/m_{d_i}^{(0)}$. With our approximations $\delta Z_{ij}^{L,R}$ is non-negligible only for $i = 3$ or $j = 3$.

In order to resum $\tan\beta$-enhanced contributions to $\delta Z_{ij}^{L,R}$ to all orders, we proceed in an anal-

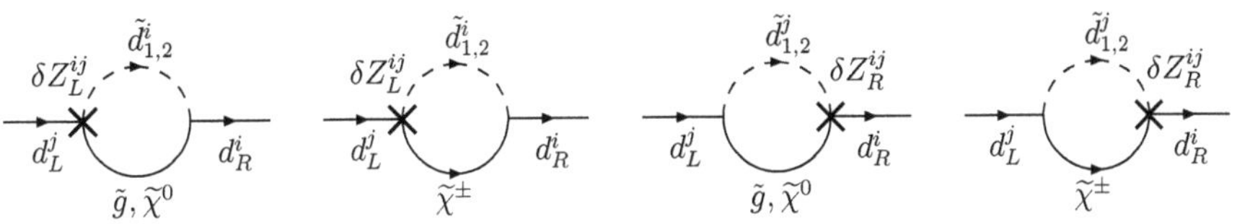

Figure 4.13: $\tan\beta$-enhanced higher-order contributions to Σ_{ij}^{RL}, generated by insertions of the counterterms δZ_{ij}^{RL} into enhanced one-loop diagrams.

ogous way as in the cases of δy_b and δV_{ij} discussed in Sections 4.3.2 and 4.4.4, respectively: We consider counterterm insertions of $\delta Z_{ij}^{L,R}$ into enhanced one-loop diagrams. The $\delta Z_{ij}^{L,R}$ enter the chargino-quark-squark, gluino-quark-squark and neutralino-quark-squark couplings thereby rendering the latter two vertices flavour non-diagonal. Therefore, beyond the one-loop level, not only chargino-squark loops contribute to Σ_{ij}^{RL} but all the diagrams shown in Fig. 4.3 for the flavour-conserving case develop flavour-changing counterparts (see Fig. 4.13). In these diagrams the flavour-change is sourced out into the wave-function counterterm and so the underlying self-energy becomes flavour-diagonal[9]. Including these additional $\tan\beta$-enhanced corrections, the self-energy Σ_{ij}^{RL} as a function of $\delta Z_{ij}^{L,R}$ reads

$$\Sigma_{ij}^{RL}(\delta Z_{ij}^L, \delta Z_{ij}^R) = V_{ti}^{(0)*} V_{tj}^{(0)} m_{d_i}^{(0)} \epsilon_{\rm FC} \tan\beta + \frac{\delta Z_{ij}^L}{2} m_{d_i}^{(0)} \epsilon_i \tan\beta - \frac{\delta Z_{ij}^R}{2} m_{d_j}^{(0)} \epsilon_j \tan\beta. \quad (4.60)$$

By writing $V^{(0)} = V + \delta V$ we allow for potential $\tan\beta$-enhanced corrections to the CKM elements V_{ij} and we know from our analysis in Section 4.4.4 that such corrections indeed will arise. Expressing Σ_{ij}^{RL} and Σ_{ij}^{LR} in Eq. (4.59) through (4.60) we find a system of equations for $\delta Z_{ij}^{L,R}$ which are valid at leading order in $\tan\beta$ but to all orders in the perturbative series. Again, they can be solved either order-by-order performing an explicit resummation or simply by solving the coupled equations directly for the resummed counterterms $\delta Z_{ij}^{L,R}$. To leading order in m_{d_i}/m_b $(i = d, s)$ the result is given by

$$\begin{aligned} \frac{\delta Z_{bi}^L}{2} &= -\frac{\epsilon_{\rm FC} \tan\beta}{1+\epsilon_b \tan\beta} V_{tb}^{(0)*} V_{ti}^{(0)}, \\ \frac{\delta Z_{bi}^R}{2} &= -\frac{m_{d_i}}{m_b} \left[\frac{\epsilon_{\rm FC} \tan\beta}{1+\epsilon_b \tan\beta} + \frac{\epsilon_{\rm FC}^* \tan\beta}{(1+\epsilon_i^* \tan\beta)} \right] V_{tb}^{(0)*} V_{ti}^{(0)} \end{aligned} \quad (4.61)$$

with all other $\delta Z_{ij}^{L,R}$ being zero for $i > j$.

Of course, the wave-function counterterms δZ_{kj}^L enter also the u_i-d_j-W^+ coupling. Therefore, to

[9]Diagrams involving a further flavour change due to a second $\delta Z^{L,R}$ insertion or an additional CKM-induced flavour-changing vertex are suppressed. This is because in Σ_{ij}^{RL} as well as in $\delta Z_{ij}^{L,R}$ only $1 \to 3$ and $2 \to 3$ transitions are non-negligible and the combination of two such transitions involves more powers of the Wolfenstein parameter λ than the LO contribution.

be able to extract the CKM matrix elements from a low energy measurement of this coupling, the δZ^L_{kj} insertions must be subtracted by appropriate CKM counterterms δV_{ij}. In addition one has to subtract also simultaneous $\delta V_{ik} \cdot \delta Z^L_{kj}$ insertions in order to account for higher order $\tan\beta$-enhanced effects. This leads to the renormalisation prescription

$$\delta V_{ij} = -\sum_k (V_{ik} + \delta V_{ik}) \frac{\delta Z^L_{kj}}{2}. \tag{4.62}$$

Solving this system of equations for the δV_{ij} we rediscover the result (4.53) obtained with the external leg approach. Inserting (4.53) into (4.61) eventually yields our final result for the wave-function counterterms:

$$\begin{aligned} \frac{\delta Z^L_{bi}}{2} &= -V^*_{tb} V_{ti} \frac{\epsilon_{\rm FC} \tan\beta}{1 + (\epsilon_b - \epsilon_{\rm FC}) \tan\beta}, \\ \frac{\delta Z^R_{bi}}{2} &= -V^*_{tb} V_{ti} \frac{m_{d_i}}{m_b} \left[\frac{\epsilon_{\rm FC} \tan\beta}{1 + \epsilon_b \tan\beta} + \frac{\epsilon^*_{\rm FC} \tan\beta}{(1 + \epsilon^*_i \tan\beta)} \right] \frac{1 + \epsilon_b \tan\beta}{1 + (\epsilon_b - \epsilon_{\rm FC}) \tan\beta}. \end{aligned} \tag{4.63}$$

4.5.2. Formulation of effective Feynman rules

In Sections 4.3.2, 4.4.4 and 4.5.1 we have determined the $\tan\beta$-enhanced counterterms δy_b, δV_{ij} and $\delta Z^{L,R}_{ij}$ to all orders in perturbation theory. Their effects can easily be incorporated into LO calculations: Instead of treating these counterterms as NLO one counts them as LO and includes them into tree-level vertices. The resulting modified Feynman rules are valid beyond the decoupling limit and refer to input scheme (i) for the sbottom parameters specified in Section 4.3.3. They can be used for calculations of low-energy processes involving virtual SUSY particles as well as for calculations in collider physics with external SUSY particles. The modifications, which can easily be implemented into computer programs like FeynArts [28], are given as follows:

(i) Express the Feynman rules in terms of the down-type Yukawa couplings y_{d_i} and replace them according to relation (4.24) by

$$y_{d_i} \to y^{(0)}_{d_i} = \frac{m_{d_i}}{v_d(1 + \epsilon_i \tan\beta)}. \tag{4.64}$$

It should be stressed that the same replacement has to be performed for the Yukawa coupling appearing in the sbottom mass matrix $M_{\tilde b}$ in (A.1) before determining the mixing angle via (A.7). In case one wants to rely on input scheme (iii) the sbottom mixing matrix has to be calculated iteratively as described in Section 4.3.3.

(ii) Replace CKM-elements involving the third quark generation according to

$$V_{ti} \longrightarrow V_{ti}^{(0)} = \frac{1+\epsilon_b \tan\beta}{1+(\epsilon_b - \epsilon_{\rm FC})\tan\beta} V_{ti} \qquad (i=d,s),$$
$$V_{ib} \longrightarrow V_{ib}^{(0)} = \frac{1+\epsilon_b^* \tan\beta}{1+(\epsilon_b^* - \epsilon_{FC}^*)\tan\beta} V_{ib} \qquad (i=u,c). \tag{4.65}$$

All other CKM-elements remain unchanged. The V_{ij} appearing after these replacements correspond to the physical ones which can be measured from the $W^+u_id_j$-vertex.

(iii) This last rule concerns vertices involving down-type quarks. Into these vertices one has to include the flavour-changing wave-function counterterms

$$\frac{\delta Z_{bi}^L}{2} = -\frac{\delta Z_{ib}^{L*}}{2} = -\frac{\epsilon_{\rm FC}\tan\beta}{1+\epsilon_b\tan\beta} V_{tb}^* V_{ti}^{(0)},$$
$$\frac{\delta Z_{bi}^R}{2} = -\frac{\delta Z_{ib}^{R*}}{2} = -\frac{m_i}{m_b}\left[\frac{\epsilon_{\rm FC}\tan\beta}{1+\epsilon_b\tan\beta} + \frac{\epsilon_{FC}^*\tan\beta}{1+\epsilon_i^*\tan\beta}\right] V_{tb}^* V_{ti}^{(0)} \tag{4.66}$$

for $i=d,s$. This leads to additional flavour-changing vertices.

If one uses our Feynman rules, $\tan\beta$-enhanced loop corrections of the form $(\epsilon\tan\beta)^n$ are automatically resumed to all orders. There is one exception: Proper vertex-corrections to the $\tan\beta$-suppressed $h^0 d^i d^j$- and $H^+ d_L^i u_R^j$-vertices and to the corresponding Goldstone-boson vertices can not be accounted for by this method.

Since we used the appropriate decoupling schemes in determining the counterterms, the SM vertices do not receive non-decoupling contributions from this replacement rules. The couplings of quarks to gauge bosons are not affected at all because the counterterms $\delta Z_{ij}^{L,R}$ cancel from the couplings to neutral gauge bosons due to their anti-hermiticity and they are removed from the W-boson vertex by the CKM renormalisation. In the couplings of the quarks to the SM-like Higgs boson h^0 and to the Goldstone bosons $G^0, G^\pm$ the non-decoupling part of the $\delta Z_{ij}^{L,R}$ and δV_{ij} insertions cancel with the non-decoupling part of the proper vertex correction which has been calculated in Ref. [21]. However, vertices which couple down quarks to the neutral Higgs bosons A^0, H^0, the charged Higgs boson H^+ or to gauginos $\tilde{g}, \tilde{\chi}^\pm, \tilde{\chi}^0$ receive flavour-changing modifications. Explicit Feynman rules for these vertices are given in Appendix A.3.

It seems that there is one type of $\delta Z_{ij}^{L,R}$-induced corrections which is not accounted for by our effective Feynman rules: The counterterms $\delta Z_{ij}^{L,R}$ enter also the down-quark mass term in the Lagrangian leading to the counterterm vertex

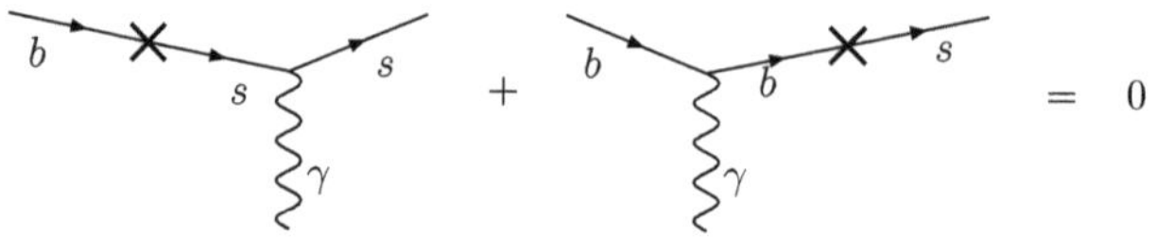

Figure 4.14: Contributions of flavour-changing wave-function counterterms in external quark legs to $b \to s\gamma$ cancel each other

$$-i\left(\frac{m_i}{1+\epsilon_i\tan\beta}\frac{\delta Z_{ij}^L}{2} - \frac{m_j}{1+\epsilon_j\tan\beta}\frac{\delta Z_{ij}^R}{2}\right)P_L$$
$$-i\left(\frac{m_i}{1+\epsilon_i^*\tan\beta}\frac{\delta Z_{ij}^R}{2} - \frac{m_j}{1+\epsilon_j^*\tan\beta}\frac{\delta Z_{ij}^L}{2}\right)P_R. \quad (4.67)$$

The inclusion of its effects into automatic calculations would be more involved since it does not simply amount to a redefinition of a tree-level vertex. However, we are not aware of a physical application where this counterterm is needed. The situation encountered here is similar to the one discussed in Section 4.3.4: The flavour-changing counterparts of the diagrams in Fig. 4.4 cancel each other in the low momentum region and survive only in the high-momentum region where the remnant is not $\tan\beta$-enhanced. We only have to worry about processes for which higher orders of the momentum expansion in m_b/M_{SUSY} are relevant because the counterterms $\delta Z_{ij}^{L,R}$ are determined at $\not{p} = 0$ and cancel only the zeroth order term in the expansion of the self-energy. An example for such a process is once again $b \to s\gamma$. However, we are lucky: The insertions of wave-function counterterms into the external quark legs (Fig. 4.14) cancel each other in this case and need not be considered.

5. PHENOMENOLOGY: RARE NON-LEPTONIC B DECAYS BEYOND THE DECOUPLING LIMIT

In the previous chapter we have discussed in detail the issue of $\tan\beta$-enhanced loop corrections to generic transition amplitudes. We arrived at a set of effective Feynman rules which allow to include the all-order resummed corrections into practical calculations. In this chapter we use these Feynman rules to determine gluino contributions to the effective Hamiltonian $\mathcal{H}_{\text{eff}}^{(1)}$ in Eq. (2.1) which is responsible for the description of rare non-leptonic B decays. These effects emerge from the fact that our treatment of the $\tan\beta$-enhanced corrections goes beyond the decoupling limit.

The most fundamental modifications to a naive LO $\mathcal{H}_{\text{eff}}^{(1)}$ arise from the additional FCNC couplings which are induced by the $\tan\beta$-enhanced wave-function counterterms $\delta Z_{ij}^{L,R}$ in (4.63). In Section 5.1 we will discuss the properties of these couplings. Due to their presence, flavour-changing transitions are no longer mediated exclusively by W bosons, charged Higgs particles and charginos but also by neutral Higgs particles, gluinos and neutralinos. For the case of the neutral Higgs bosons, the phenomenological effects on rare B decays, especially on the decay $B_s \to \mu^+\mu^-$, have widely been studied in the framework of the effective 2HDM valid for $M_{\text{SUSY}} \gg v$ [17, 18, 20, 49]. With this method it is, however, not possible to assess the FCNC gluino- and neutralino- couplings and to calculate analytic formulae for their contributions to the Wilson coefficients C_i. Our effective Feynman rules, on the other hand, enable us to perform such a calculation. The results for the gluino contributions are collected in Appendix A.4 and their importance is discussed in Section 5.2. We found that the δZ_{ij}^{L}-induced gluino-squark loops have a large impact on the Wilson coefficient C_{8g} of the chromomagnetic operator. To illustrate the phenomenological consequences of this gluino-squark contribution, we discuss its effect on the mixing induced CP asymmetry in the decay $B^0 \to \phi K_s$ in Section 5.3.

5.1. Effective FCNC couplings

The FCNC couplings generated by the counterterms δZ_{bi}^{L} and δZ_{bi}^{R} $(i = d, s)$ originate from $\tan\beta$-enhanced flavour-changing self-energies. Therefore their numerical importance crucially depends on the parameter $\epsilon_{\text{FC}} \tan\beta$. Since δZ_{bi}^{R} is suppressed by a small ratio of quark masses,

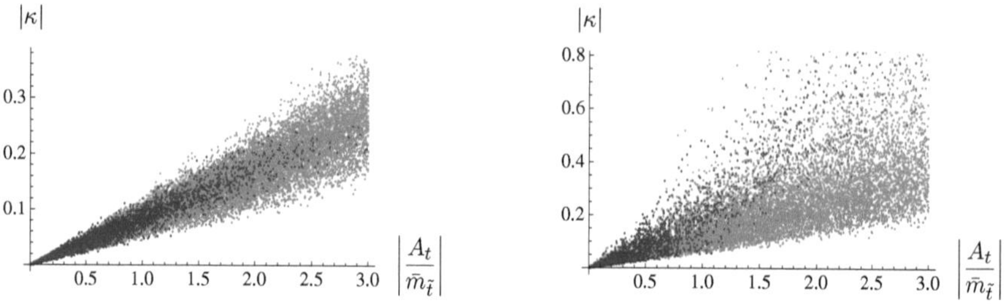

Figure 5.1: Size of the coupling κ as a function of $|A_t/\bar{m}_{\tilde{t}}|$ for a scan over the MSSM parameter space with positive μ (left) and negative μ (right). The orange points are eliminated by a later application of the constraints from $B \to X_s\gamma$.

the most important effects come from δZ^L_{bi} in (4.63). The resulting FCNC couplings thus have the form

$$\kappa_{bi} \propto \kappa \cdot V^*_{tb} V_{ti}, \qquad \text{with} \quad \kappa = \frac{\epsilon_{\rm FC} \tan\beta}{1 + (\epsilon_b - \epsilon_{\rm FC}) \tan\beta}. \tag{5.1}$$

We see that the effective coupling κ_{bi} preserves the CKM structure of MFV and its strength is given by the parameter combination κ. It is thus useful to have a first estimate of the size of κ. For this purpose, we neglect the weak contributions to ϵ_b and $\epsilon_{\rm FC}$, focus on the non-decoupling part of expressions (4.16) and (4.37) for $\epsilon_b^{\tilde{g}}$ and $\epsilon_{\rm FC}$ and set all the SUSY mass parameters as well as $|\mu|$ and $|A_t|$ equal to a single mass scale $M_{\rm SUSY}$. In this case, the mass dependence drops out and we obtain

$$\begin{aligned} |\epsilon_{\rm FC} \tan\beta| &= \frac{y_t^2}{32\pi^2} \tan\beta, \\ |(\epsilon_b - \epsilon_{\rm FC}) \tan\beta| &= |\epsilon_b^{\tilde{g}} \tan\beta| = \frac{\alpha_s}{3\pi} \tan\beta \end{aligned} \tag{5.2}$$

with $y_t = y_t(M_{\rm SUSY})$ and $\alpha_s = \alpha_s(M_{\rm SUSY})$. For $\tan\beta = 50$ and $M_{\rm SUSY} = 500$ GeV, we find typical numerical values of

$$|\epsilon_{\rm FC} \tan\beta| \sim 0.12, \qquad |(\epsilon_b - \epsilon_{\rm FC}) \tan\beta| \sim 0.5\,. \tag{5.3}$$

Taking μ real here the parameter κ is evaluated as

$$|\kappa| \sim 0.08, \quad \text{if}\ \mu > 0, \qquad |\kappa| \sim 0.24, \quad \text{if}\ \mu < 0. \tag{5.4}$$

Values larger than this for $\epsilon_{\rm FC}$ and thus for $|\kappa|$ occur if $|A_t|$ is significantly larger than the masses of stops and charginos. If one requires $|A_t| \lesssim 3\bar{m}_{\tilde{t}}$ (where $\bar{m}_{\tilde{t}} = \sqrt{\tilde{m}_Q \tilde{m}_t}$ is an average stop mass) to avoid colour-breaking minima [69], $\epsilon_{\rm FC} \tan\beta$ gets constrained to $|\epsilon_{\rm FC} \tan\beta|_{\rm max} \sim 0.4$.

Experimentally, the size of A_t is further limited by $\mathrm{Br}(\bar{B} \to X_s\gamma)$ via the $\tan\beta$-enhanced chargino contribution to this process. Note however that this bound might significantly be shifted when the new contributions, e.g. from gluino-squark loops, are taken into account. Therefore the $\text{Br}(\bar{B} \to X_s\gamma)$ constraint can only be applied *after* inclusion of the κ_{bs}-induced contributions, it cannot be used to constrain the size of this coupling *a priori*.

We stress that complex values of A_t lead to complex κ and hence to additional CP violation in the FCNC couplings κ_{bi}. Assuming complex values for A_t has the additional advantage that the bound from $\bar{B} \to X_s\gamma$ on $|A_t|$ is much weaker in this case [70] and one can have a larger coupling strength $|\kappa|$.

In Fig. 5.1 we show the coupling strength $|\kappa|$ for a scan over the relevant MSSM parameter space for positive μ (left plot) and negative μ (right plot). To this end we scanned the mass parameters $\widetilde{m}_{Q,t,b}$, $|\mu|$, $M_{1,2}$ and $m_{\tilde{g}}$ from 200 GeV to 1000 GeV and $\tan\beta$ from 40 to 60. The absolute value of A_t is varied between 0 and $3\widetilde{m}_{\tilde{t}}$ and attributed by an arbitrary phase φ_{A_t}. Only parameter points compatible with the following constraints have been accepted:

- All squark masses are larger than 200 GeV.
- The lightest supersymmetric particle (LSP) is charge- and color-neutral.
- The experimental 2σ-bound on the lightest Higgs-boson mass is respected.

The light-coloured points fulfill these conditions. We see that $|\kappa|$ can a priori reach values up to $|\kappa| \lesssim 0.4$ for positive μ and values up to $|\kappa| \lesssim 1.0$ (and even larger) for negative μ. After application of constraints from $B \to X_s\gamma$, as discussed in the following section, the allowed region shrinks to the dark-coloured points with the consequence of $|\kappa|$ being limited for positive μ to $|\kappa| \lesssim 0.2$.

5.2. Gluino contributions to the effective $\Delta B = 1$ Hamiltonian

Rare non-leptonic B decays are described by the effective Hamiltonian (2.1). In the SM it receives contributions from penguin and box diagrams with the $b \to s$ FCNC being mediated by a W-top loop. In the MSSM further contributions arise from the same topological diagrams with the FCNC mediated instead by SUSY particles. Under the assumption of naive MFV, only chargino and charged Higgs diagrams appear at one loop. However, an improved LO calculation should include $\tan\beta$-enhanced higher order corrections and at this point additional contributions involving gluinos, neutralinos or neutral Higgs bosons enter. These corrections can be accounted for by performing an ordinary one-loop calculation but using the effective Feynman rules of Section 4.5.2. This amounts in treating the couplings κ_{bi}, discussed in the previous section, as tree-level.

Neutral Higgs diagrams have previously been studied using the effective theory approach. In particular the corresponding coefficients $C_{7\gamma}^{H^0}$ and $C_{8g}^{H^0}$ of the magnetic and chromomagnetic operators have been presented in Ref. [7]. Analytic results for the gluino and neutralino coefficients, on the other hand, cannot be obtained in this way but they can be calculated with our diagrammatic method. We will concentrate on the gluino contributions since they are expected to dominate over the neutralino ones due to the strong coupling. We calculated the whole set of gluino diagrams including gluon, photon and Z penguins as well as box diagrams. The resulting Wilson coefficients are given in Appendix A.4.

In Section 3.2.3 we argued that largest modifications in the MSSM with MFV are expected for the Wilson coefficients of the magnetic and chromomagnetic operators which allow for a potential $\tan\beta$-enhancement with respect to the SM coefficient. This is indeed a property of the well-known chargino-contribution for which we find

$$\begin{aligned} C_{7\gamma,8g}^{\tilde{\chi}^\pm} &= \frac{1}{\cos\beta(1+\epsilon_b^*\tan\beta)} \sum_{a=1,2} \Bigg\{ \frac{\widetilde{U}_{a2}\widetilde{V}_{a1}\sqrt{2}M_W}{m_{\tilde{\chi}_a^\pm}} \left[K^* g_{7\gamma,8g}(x_{\tilde{q}\tilde{\chi}_a^\pm}) - c_{\tilde{t}}^2\, g_{7\gamma,8g}(x_{\tilde{t}_1\tilde{\chi}_a^\pm})\right. \\ &\left. - s_{\tilde{t}}^2\, g_{7\gamma,8g}(x_{\tilde{t}_2\tilde{\chi}_a^\pm})\right] + s_{\tilde{t}}\, c_{\tilde{t}}\, e^{-i\phi_{\tilde{t}}} \frac{\widetilde{U}_{a2}\, \widetilde{V}_{a2}\, m_t}{\sin\beta\, m_{\tilde{\chi}_a^\pm}} \left[g_{7\gamma,8g}(x_{\tilde{t}_1\tilde{\chi}_a^\pm}) - g_{7\gamma,8g}(x_{\tilde{t}_2\tilde{\chi}_a^\pm})\right] \Bigg\}. \end{aligned} \quad (5.5)$$

with

$$s_{\tilde{t}} = \sin\tilde{\theta}_t, \qquad c_{\tilde{t}} = \cos\tilde{\theta}_t, \qquad x_{ij} = m_i/m_j. \quad (5.6)$$

Here $m_{\tilde{q}}$ denotes again the common mass of left-handed squarks of the first two generations. All loop functions are given in Appendix A.2. Our result differs from the one in [21] only by a factor of K^* (defined in (4.53)) in the numerically small up and charm squark contribution. The stop contribution remains unaffected because the corrections from the wave function and the CKM counterterm cancel each other.

Also the novel gluino contribution develops a $\tan\beta$-enhanced part which reads

$$\begin{aligned} C_{7\gamma,8g}^{\tilde{g}} &= \frac{\sqrt{2}}{4G_F} \frac{g_s^2\mu\tan\beta}{m_{\tilde{g}}(m_{\tilde{b}_1}^2 - m_{\tilde{b}_2}^2)} \frac{\epsilon_{\rm FC}^*\tan\beta}{(1+\epsilon_b^*\tan\beta)\,(1+(\epsilon_b^*-\epsilon_{\rm FC}^*)\tan\beta)} \\ &\times \left[C_F\left(f_{7\gamma,8g}^F(x_{\tilde{b}_1\tilde{g}}) - f_{7\gamma,8g}^F(x_{\tilde{b}_2\tilde{g}})\right) + C_A\left(f_{7\gamma,8g}^A(x_{\tilde{b}_1\tilde{g}}) - f_{7\gamma,8g}^A(x_{\tilde{b}_2\tilde{g}})\right)\right] \end{aligned} \quad (5.7)$$

with the colour factors $C_F = 4/3$ and $C_A = 3$. Note that this contribution is of the same order $\mathcal{O}(y_t^2\,\alpha_s\,\tan^2\beta \times v^2/M_{\rm SUSY}^2)$ (plus resummed higher orders) as the ϵ_b correction in the chargino terms (5.5). While the latter is usually included in LO calculations, the former has not been discussed yet in the literature since the effective theory method commonly used for the resummation gives no handle on it. Of course, the gluino effects are contained in the numerical 2-loop calculation performed in Ref. [71]. New features of our result are the identification of the $\tan\beta$-

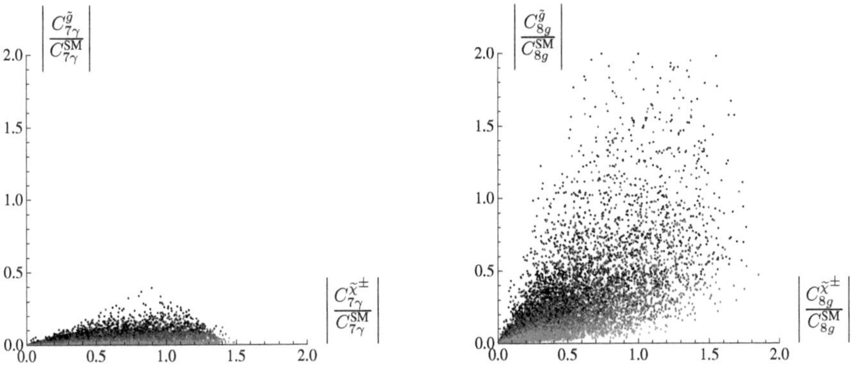

Figure 5.2: Chargino- and gluino-contributions to the Wilson coefficients $C_{7\gamma}$ (left) and C_{8g} (right) for a scan over the MSSM parameter space. Different colours correspond to different ranges of values for μ: light: $200\,\mathrm{GeV} < \mu < 600\,\mathrm{GeV}$, dark: $600\,\mathrm{GeV} < \mu < 1000\,\mathrm{GeV}$.

enhanced part of the 2-loop contributions, the determination of an analytic expression for it and the resummation of $\tan\beta$-enhanced corrections to all orders beyond the decoupling limit.

To have a rough estimate of the size of $C^{\tilde{g}}_{7\gamma,8g}$ compared to $C^{\tilde{\chi}^\pm}_{7\gamma,8g}$ we again set all SUSY masses (including $|\mu|$ and $|A_t|$) to the same value M_{SUSY}. In this case we find

$$\eta_{7\gamma} = \left|\frac{C^{\tilde{g}}_{7\gamma}}{C^{\tilde{\chi}^\pm}_{7\gamma}}\right| = \frac{8}{15}\frac{g_s^2}{y_t^2}|\kappa|\,, \qquad \eta_{8g} = \left|\frac{C^{\tilde{g}}_{8g}}{C^{\tilde{\chi}^\pm}_{8g}}\right| = \frac{10}{3}\frac{g_s^2}{y_t^2}|\kappa|\,. \tag{5.8}$$

Using our estimates (5.4) for $|\kappa|$ we find $\eta_{7\gamma} \sim 0.07$ and $\eta_{8g} \sim 0.42$ for positive values of μ and $\eta_{7\gamma} \sim 0.2$ and $\eta_{8g} \sim 1.3$ for negative values of μ. It follows that the impact of the gluino contribution on $C_{7\gamma}$ is small (especially for positive μ) whereas the contribution to C_{8g} can be sizeable. In the previous section we argued that the value of $|\kappa|$ can be increased if one chooses large values for $|A_t|$. Of course, the size of $C^{\tilde{g}}_{7\gamma,8g}$ gets larger for increasing values of $|\kappa|$. Note, however, that $C^{\tilde{\chi}^\pm}_{7\gamma,8g}$ is proportional to A_t and thus the ratio $\eta_{7\gamma,8g}$, i.e. the relative importance of the gluino contribution, is essentially unaffected. On the other hand, the gluino contribution grows with increasing $|\mu|$ whereas the chargino contribution decreases because it decouples with the chargino mass. Therefore for large values of $|\mu|$ the gluino contribution becomes more important.

A more detailed numerical study of the coefficients $C^{\tilde{\chi}^\pm\tilde{g}}_{7\gamma,8g}$ is shown in Fig. 5.2. There the values of these coefficients at the scale m_t, normalised to the corresponding SM coefficients, are displayed for a scan of the MSSM parameter space. We have chosen μ positive and otherwise used the same range of values and constraints as in Section 5.1. Furthermore with the full $C_{7\gamma,8g}$ at hand we are in the position to apply additional constraints from $\bar{B} \to X_s\gamma$:

- The inclusive decay $\bar{B} \to X_s\gamma$ is at tree-level mediated by the magnetic operator $Q_{7\gamma}$[1]. Therefore the experimental value [42]

$$\text{Br}(\bar{B} \to X_s\gamma) \overset{\text{exp.}}{=} (3.55 \pm 0.24^{+0.09}_{-0.10} \pm 0.03) \times 10^{-4} \tag{5.9}$$

  is highly sensitive to the coefficient $C_{7\gamma}$. The SM value of the branching fraction has been calculated at NNLO and reads [72]

$$\text{Br}(\bar{B} \to X_s\gamma) \overset{\text{SM}}{=} (3.15 \pm 0.23) \times 10^{-4}. \tag{5.10}$$

  Adding errors in quadrature we find for the ratio of experimental and theory result

$$R \equiv \frac{\text{Br}(\bar{B} \to X_s\gamma)^{\text{exp}}}{\text{Br}(\bar{B} \to X_s\gamma)^{\text{SM}}} = 1.13 \pm 0.12. \tag{5.11}$$

  For our $\bar{B} \to X_s\gamma$ constraint we evaluate the quantity R replacing $\text{Br}(\bar{B} \to X_s\gamma)^{\text{exp}}$ by our MSSM theory value and $\text{Br}(\bar{B} \to X_s\gamma)^{\text{SM}}$ by our less accurate SM value, both calculated according to Eq. (20) of Ref. [73]. Then equality with the result (5.11) is demanded at the $2\,\sigma$ level. In the MSSM branching ratio we include also the (small) neutralino- as well as the (not $\tan\beta$-enhanced) charged Higgs contribution.

- The fact that we consider complex A_t opens up the possibility to fulfill the $\bar{B} \to X_s\gamma$ constraint by fine-tuning the phase φ_{A_t} of A_t. In order to avoid such an unnatural situation we impose the additional condition $|C_{7\gamma}^{\text{SUSY}}| \leq |C_{7\gamma}^{\text{SM}}|$.

The results of the scan in Fig. 5.2 confirm our rough estimate in Eq. (5.8) and the discussion below this equation: Whereas the gluino contribution does not significantly affect $C_{7\gamma}$, its effect on C_{8g} can be sizable and increases with $|\mu|$.

SUSY contributions to other Wilson coefficients lack the $\tan\beta$-enhancement and are thus less important. The novel gluino effects are even much smaller than the chargino ones because they suffer from GIM-like cancellations. From the results in Appendix A.4 we see that their contributions to gluon or photon penguins as well as to box diagrams are proportional to the structure

$$\left(f(x_{\tilde{b}_1\tilde{g}}) - f(x_{\tilde{q}\tilde{g}})\right) + \widetilde{R}^{b*}_{21}\widetilde{R}^{b}_{21}\left(f(x_{\tilde{b}_2\tilde{g}}) - f(x_{\tilde{b}_1\tilde{g}})\right) \tag{5.12}$$

where f is the respective loop function. This expression exhibits a v^2/M^2_{SUSY} suppression since $m^2_{\tilde{b}_1} = m^2_{\tilde{q}} + \mathcal{O}(v^2/M^2_{\text{SUSY}})$ and $\widetilde{R}^b_{21} = \mathcal{O}(v/M_{\text{SUSY}})$. Moreover, also the v^2/M^2_{SUSY}-terms cancel for $m_{\tilde{b}_L} = m_{\tilde{b}_R}$ as it is often assumed or approximately fulfilled like in the popular mSUGRA

[1] One-loop contributions of the operators $Q_1, ..., Q_6$ which are comparable in size are absorbed into an effective redefinition of the operator $Q_{7\gamma}$

scenarios. In combination with the mandatory $v^2/M_{\rm SUSY}^2$ factor of a SUSY coefficient to a dimension six operator we have a total

$$v^4/M_{\rm SUSY}^4 \times \left(v^2/M_{\rm SUSY}^2 \quad \text{or} \quad (m_{\tilde{b}_L}^2 - m_{\tilde{b}_R}^2)/M_{\rm SUSY}^2\right) \tag{5.13}$$

suppression. Even going beyond the decoupling limit, one has to choose at least $m_{\tilde{b}_{L,R}} \gtrsim \sqrt{2}v$ in order to avoid that the mass splitting induced by the offdiagonal elements of the mass matrix drives the lighter sbottom mass below our 200 GeV limit. Origin of the GIM-like suppression is the $U(3)$-flavour symmetry, respected by the strong coupling and the left-handed SUSY breaking mass term. Such a suppression is therefore absent for the chargino contribution because the chargino coupling breaks this symmetry.

The GIM-like cancellation does not occur in diagrams which require a chirality-flip breaking the $U(3)$ flavour symmetry. This was the case for the magnetic and chromomagnetic operators and it also happens in case of the Z penguin. Here the gluino contribution is proportional to

$$\widetilde{R}_{i1}^{b*}\,\widetilde{R}_{i2}^{b}\,\widetilde{R}_{j2}^{b*}\,\widetilde{R}_{j1}^{b}\,f_Z(x_{\tilde{b}_i\tilde{g}}, x_{\tilde{b}_2\tilde{g}}) \tag{5.14}$$

with two chirality flips encoded in the mixing matrices. Since the $\tilde{b}_{1,2}$ contributions tend to cancel each other due to the unitarity of $\widetilde{R}^b$, the structure of the gluino Z penguin contribution is

$$v^2/M_{\rm SUSY}^2 \times \left(v^2/M_{\rm SUSY}^2 \quad \text{or} \quad (m_{\tilde{b}_L}^2 - m_{\tilde{b}_R}^2)/M_{\rm SUSY}^2\right)^2. \tag{5.15}$$

This prevents the gluino contribution again from competing with the chargino one and even more with the SM one which receives a m_t^2/M_W^2 enhancement.

In summary, gluino diagrams are only relevant for the magnetic and chromomagnetic operators $Q_{7\gamma,8g}$. Whereas the gluino contribution to $C_{7\gamma}$ is accidentally suppressed, it is enhanced for C_{8g} and can yield sizable corrections, especially for large values of $|\mu|$.

5.3. The mixing-induced CP asymmetry in $\bar{B}^0 \to \phi K_s$

The gluino contribution to C_{8g} affects some important low-energy observables in non-leptonic B decays. As an example, we study its impact on the mixing-induced CP asymmetry in the decay $\bar{B}^0 \to \phi K_s$. NP contributions to CP asymmetries of hadronic B^0 decays have been analysed in a model-independent way in Ref. [74]. Among other scenarios, also the case of a dominant NP coefficient $C_{8g}^{\rm NP}$ has been considered using an improved LO framework of QCDF. In Ref. [75] the result has been applied to the flavour-blind MSSM with complex A_t and a large effect in the mixing-induced CP asymmetry in the decay $\bar{B}^0 \to \phi K_s$ has been found. Here we improve these

a_f^u	$0.02^{+0.01}_{-0.01} + 0.005^{+0.013}_{-0.002}\, i$	$b_{f,6}^c$	$(-30)^{+65}_{-6} + 8^{+44}_{-52}\, i$
$b_{f,1}^c$	$1.13^{+0.31}_{-0.05} + 0.02^{+0.18}_{-0.26}\, i$	$b_{f,7}^c$	$15^{+10}_{-3} + (-0.1)^{4.3}_{-8.9}\, i$
$b_{f,1}^u$	$0.02^{+0.02}_{-0.01} + 0.010^{0.024}_{-0.004}\, i$	$b_{f,8}^c$	$20^{+3}_{-30} + (-3)^{+25}_{-20}\, i$
$b_{f,2}^c$	$(-0.38)^{+0.07}_{-0.56} + (-0.05)^{+0.46}_{0.34}\, i$	$b_{f,9}^c$	$11^{+26}_{-4} + 2^{+15}_{-23}\, i$
$b_{f,2}^u$	$(-0.008)^{+0.005}_{-0.014} + 0.003^{+0.014}_{-0.003}\, i$	$b_{f,10}^c$	$12^{+27}_{-4} + 2^{+15}_{-23}\, i$
$b_{f,3}^c$	$(-24)^{+8}_{-54} + (-3)^{+47}_{-30}\, i$	$b_{f,7\gamma}^c$	$(-0.02)^{+0.01}_{-0.04} + (-0.003)^{+0.029}_{-0.021}\, i$
$b_{f,4}^c$	$(-18)^{+8}_{-54} + (-3)^{+47}_{-30}\, i$	$b_{f,8g}^c$	$0.60^{+1.45}_{-0.25} + 0.11^{+0.86}_{-1.21}\, i$
$b_{f,5}^c$	$(-28)^{+6}_{-19} + 1^{+17}_{-8}\, i$		

Table 5.1: Numerical values of a_f^u and of the $b_{f,i}^{u,c}$ $(i = 1, \ldots, 10, 7\gamma, 8g)$ parameters at NLO QCDF. For $i = 3, .., 10, 7\gamma, 8g$ one has $b_{f,i}^u = \left(\lambda_u^{(s)} / \lambda_c^{s)}\right) b_{f,i}^c$.

previous analyses using our full NLO QCDF approach and we demonstrate the importance of the novel gluino contribution.

Mixing induced CP violation in a decay of the B^0 meson into a CP eigenstate f can be accessed through a measurement of the time-dependent CP asymmetry

$$a_f(t) = \frac{\mathrm{Br}(\bar{B}^0(t) \to f) - \mathrm{Br}(B^0(t) \to f)}{\mathrm{Br}(\bar{B}^0(t) \to f) + \mathrm{Br}(B^0(t) \to f)} = C_f \cos(\Delta M_B t) + S_f \sin(\Delta M_B t) \qquad (5.16)$$

with ΔM_B denoting the mass difference of the two B^0 meson mass eigenstates. The mixing induced CP violation resides in the quantity S_f whose theory value is given in terms of the decay amplitudes $\mathcal{A}_f = \mathcal{A}(B^0 \to f)$ and $\bar{\mathcal{A}} = \mathcal{A}_f(\bar{B}^0 \to f)$ by

$$S_f = \frac{2\,\mathrm{Im}(\lambda_f)}{1 + |\lambda_f|^2} \qquad \text{with} \quad \lambda_f = -e^{-i\phi_B} \frac{\bar{\mathcal{A}}_f}{\mathcal{A}_f}. \qquad (5.17)$$

Here ϕ_B is the B^0-$\bar{B}^0$ mixing phase.

Following Ref. [74] we parametrise the decay amplitude in terms of the NP coefficients C_i^{NP} as[2]

$$\bar{\mathcal{A}}_f = \bar{\mathcal{A}}_f^c \left[1 + a_f^u e^{-i\gamma} + \sum_i \left(b_{f,i}^c + b_{f,i}^u e^{-i\gamma} \right) C_i^{\mathrm{NP}} \right], \qquad i = 1, ..., 10, 7\gamma, 8g \qquad (5.18)$$

with the CKM angle $\gamma = \arg\left[- (V_{ud} V_{ub}^*) / (V_{cd} V_{cb}^*) \right]$. In Ref. [74] the parameters a_f^u, $b_{f,i}^{u,c}$ have been calculated using naive factorisation except for $b_{f,7\gamma,8g}^{u,c}$ which vanish in this approximation. In order to study the impact of a NP contribution C_{8g}^{NP}, the parameter b_{8g}^c has been included at NLO in QCDF which is consistent if one assumes dominance of C_{8g}^{NP} over the other NP coefficients. Note that in the MSSM this condition is fulfilled because the chargino and gluino contributions to

[2] There is actually a complex conjugation of the Wilson coefficients C_i^{NP} missing in the corresponding equation in [74].

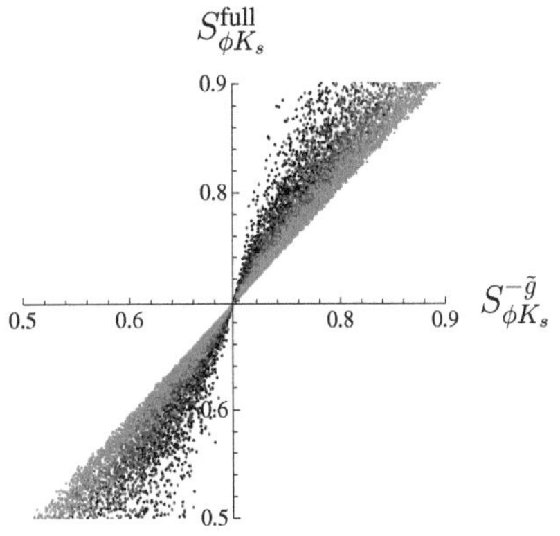

Figure 5.3: $S_{\phi K_s}$ for a scan over the MSSM parameter space: Result without the gluino contribution $S^{-\tilde{g}}_{\phi K_s}$ vs. full result $S^{\text{full}}_{\phi K_s}$.

$C_{7\gamma,8g}$ are $\tan\beta$-enhanced and effects from $C_{7\gamma}$ in hadronic B decays are $\alpha_e/\alpha_s(m_b)$ - suppressed compared to effects from C_{8g}. For our study we use the full NLO QCDF expression and present our improved values for the coefficients a^u_f, $b^{u,c}_{f,i}$ in Tab. 5.1. In our naive MFV framework we find effects from C^{NP}_3,...,C^{NP}_{10} to be negligible while the contribution from $C_{7\gamma}$ gives corrections up to 10%. Hence the approximate formulae of Ref. [74] gives reasonable results. Note, however, that our value for the relevant parameter $b^c_{f,8g}$ strongly differs from the one given in Ref. [74].

In the diagram in Fig. 5.3 the observable $S_{\phi K_S}$ is shown for a paramter scan over the MSSM parameter space covering the same ranges for the parameters and fulfilling the same constraints as discussed in the previous sections. The parameter μ is again chosen positive and its magnitude is represented by the same colour code as in Fig. 5.2. We show the result containing only the chargino- and the charged-Higgs contribution (x-axis) versus the one with the additional gluino part (y-axis). The distance of the points from the diagonal signals the importance of the gluino contribution. We see that for positive μ the gluino effect adds always constructively and that the gluino contribution can modify the result a lot. For a further illustration we have plotted $S_{\phi K_S}$ in Fig. 5.4 as a function of $|A_t|$ fixing the other parameters to certain values. The parameter point chosen for the plot fulfills all constraints mentioned above over the whole range of $|A_t|$. For the result corresponding to the dashed line the novel gluino effect has been omitted whereas the result represented by the solid line takes it into account. Also this plot demonstrates that the gluino-squark contribution can indeed have a large impact on $S_{\phi K_S}$ for complex A_t. The current

$\tilde{m}_{Q_L}, \tilde{m}_{u_R}, \tilde{m}_{d_R}$	600 GeV
μ	800 GeV
M_1	300 GeV
M_2	400 GeV
M_3	500 GeV
$\tan\beta$	50
φ_{A_t}	$3\pi/2$

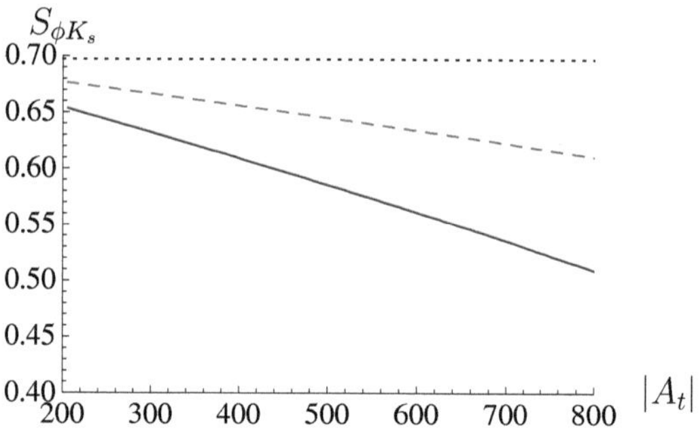

Figure 5.4: $S_{\phi K_S}$ as a function of $|A_t|$ at the parameter point shown in the table: Full result (solid) and result without the gluino contribution (dashed). The dotted line is the SM prediction.

experimental value is given by [42]

$$S_{\phi K_s} \stackrel{\text{exp.}}{=} 0.56^{+0.16}_{-0.18}\,. \tag{5.19}$$

Part II

Probing new physics in electroweak penguins via hadronic B decays

6. Isospin violation in $B \to K\pi$ decays

The $B \to K\pi$ decays are dominated by the isospin-conserving QCD penguin amplitude. Nevertheless, combined measurements of the four different decay modes $B^- \to \bar{K}^0\pi^-$, $B^- \to K^-\pi^0$, $\bar{B}^0 \to K^-\pi^+$ and $\bar{B}^0 \to \bar{K}^0\pi^0$ are sensitive to isospin violation. This is because the amplitudes are related in the limit of exact isospin symmetry via Clebsch-Gordon coefficients and any deviation from this pattern signals isospin breaking.

In this chapter we perform an isospin decomposition of the $B \to K\pi$ amplitudes and relate the resulting isospin amplitudes then to topological expressions. We then construct observables which are sensitive to isospin violation, calculate their SM values within our QCDF framework and draw conclusions on potential new physics.

6.1. Isospin decomposition of the amplitudes

The strong interaction is flavour-blind, for example a gluon cannot distinguish an up-quark from a down-quark. This means, as long as we neglect quark masses, the QCD Lagrangian is invariant under simultaneous rotations

$$q_i \to U_{ij} q_j, \qquad \bar{q}_i \to \bar{q}_j U^\dagger_{ji} = U^*_{ij}\bar{q}_j \tag{6.1}$$

of quarks and anti-quarks in flavour space. Restricting ourselves to the subspace spanned by the up- and down quarks, for which the assumption of negligible masses is clearly fulfilled, we encounter the $SU(2)$ symmetry of strong isospin. Quarks and anti-quarks transform under this $SU(2)$ as isospin-doublets[1]

$$(u, d)_{1/2}, \qquad (\bar{d}, -\bar{u})_{1/2}. \tag{6.2}$$

Since strong isospin is conserved in QCD, it serves as a good quantum number to classify mesons and baryons and it is attributed to them according to their valence quark content. The mesons

[1] Transforming $(\bar{u}, \bar{d})^T$ by the conjugate rotation U^* as in (6.1) is equivalent to transforming $(i\sigma_2)(\bar{u}, \bar{d})^T = (\bar{d}, -\bar{u})^T$ by the fundamental rotation U. This can be seen by noting that $(i\sigma_2)U^*(i\sigma_2)^\dagger = U$ for any $U \in SU(2)$.

participating in $B \to K\pi$ decays transform under isospin rotations as

$$(\bar{B}^0, -B^-)_{1/2}, \qquad (\bar{K}^0, -K^-)_{1/2}, \qquad (\pi^+, -\pi^0, -\pi^-)_1 . \tag{6.3}$$

In a similar way we can assign isospin to the operators appearing in the effective Hamiltonian (2.1) which mediates the $B \to K\pi$ transitions. Containing $\bar{u}u$- and $\bar{d}d$-pairs, the operators Q_1,...,Q_{10} can be distributed among

$$\mathcal{H}_{\text{eff}} = \mathcal{H}_{\text{eff}}^{\Delta I=0} + \mathcal{H}_{\text{eff}}^{\Delta I=1} \tag{6.4}$$

according to the decomposition $1/2 \otimes 1/2 = 1 \oplus 0$. Since the QCD penguin operators $Q_3, ..., Q_6$ involve the isosinglet combination $(\bar{u}u + \bar{d}d)$, they contribute solely to $\mathcal{H}_{\text{eff}}^{\Delta I=0}$ whereas the other operators give contributions to both parts of $\mathcal{H}_{\text{eff}}$. The $B \to K\pi$ decays thus exhibit the following isospin structure:

$$1/2 \overset{\Delta I=0,1}{\longrightarrow} 1/2 \otimes 1 = 3/2 \oplus 1/2 \tag{6.5}$$

Having specified the transformation properties of mesons and operators, we can now exploit the isospin symmetry to gain information on the decay amplitudes. Acting on the initial $I = 1/2$ B-meson states, $\mathcal{H}_{\text{eff}}$ can rise the total isospin quantum number at most by one unit conserving at the same time the z-component $I_z = \pm 1/2$. The non-vanishing matrix elements are correlated due to the Wigner-Eckart theorem. In the basis of total isospin for the final states they can be parametrised in terms of reduced matrix elements as

$$\begin{aligned}
-\sqrt{\frac{3}{2}}\,\mathcal{A}_{1/2}^{\Delta I=0} &\equiv \langle 1/2, 1/2 |\, \mathcal{H}_{\text{eff}}^{\Delta I=0} \,| \bar{B}^0 \rangle = -\langle 1/2, -1/2 |\, \mathcal{H}_{\text{eff}}^{\Delta I=0} \,| B^- \rangle, \\
-\sqrt{3}\,\mathcal{A}_{3/2}^{\Delta I=1} &\equiv \langle 3/2, 1/2 |\, \mathcal{H}_{\text{eff}}^{\Delta I=1} \,| \bar{B}^0 \rangle = -\langle 3/2, -1/2 |\, \mathcal{H}_{\text{eff}}^{\Delta I=1} \,| B^- \rangle, \\
\sqrt{\frac{3}{2}}\,\mathcal{A}_{1/2}^{\Delta I=1} &\equiv \langle 1/2, 1/2 |\, \mathcal{H}_{\text{eff}}^{\Delta I=1} \,| \bar{B}^0 \rangle = \langle 1/2, -1/2 |\, \mathcal{H}_{\text{eff}}^{\Delta I=1} \,| B^- \rangle .
\end{aligned} \tag{6.6}$$

The existence of only three independent matrix elements implies a relation linking the four $B \to K\pi$ amplitudes. To find the decay amplitudes one decomposes the $K\pi$-final states as $1/2 \otimes 1 = 3/2 \oplus 1/2$ via Clebsch-Gordon coefficients and evaluates the matrix elements with the help of (6.6). This results in

$$\begin{aligned}
\mathcal{A}(B^- \to \bar{K}^0\pi^-) &= \mathcal{A}_{1/2}^{\Delta I=0} - \mathcal{A}_{3/2}^{\Delta I=1} + \mathcal{A}_{1/2}^{\Delta I=1} \\
\sqrt{2}\,\mathcal{A}(B^- \to \bar{K}^-\pi^0) &= \mathcal{A}_{1/2}^{\Delta I=0} + 2\,\mathcal{A}_{3/2}^{\Delta I=1} + \mathcal{A}_{1/2}^{\Delta I=1} \\
\mathcal{A}(\bar{B}^0 \to K^-\pi^+) &= \mathcal{A}_{1/2}^{\Delta I=0} + \mathcal{A}_{3/2}^{\Delta I=1} - \mathcal{A}_{1/2}^{\Delta I=1} \\
\sqrt{2}\,\mathcal{A}(\bar{B}^0 \to \bar{K}^0\pi^0) &= -\mathcal{A}_{1/2}^{\Delta I=0} + 2\,\mathcal{A}_{3/2}^{\Delta I=1} + \mathcal{A}_{1/2}^{\Delta I=1},
\end{aligned} \tag{6.7}$$

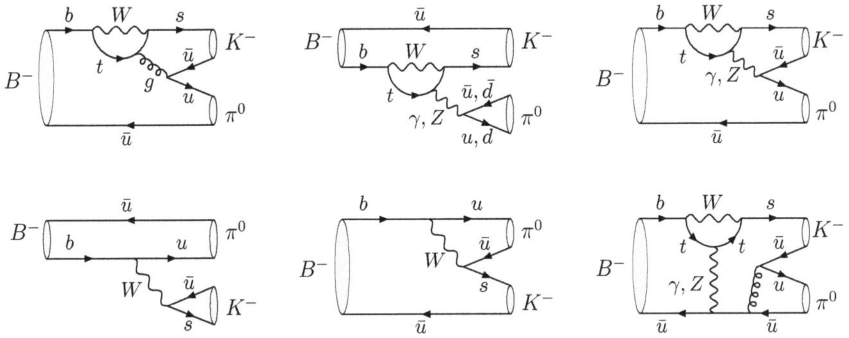

Figure 6.1: Diagrams representing the topological parametrisation in Eq. (6.9) for $B^- \to K^-\pi^0$. First line from left to right: QCD penguin (P), colour-allowed electroweak penguin ($r_{\rm EW}$), colour-suppressed electroweak penguin ($r_{\rm EW}^{\rm C}$). Second line from left to right: Colour-allowed tree ($r_{\rm T}$), colour-suppressed tree ($r_{\rm C}$), electroweak penguin annihilation ($r_{\rm EW}^{\rm A}$).

from where we can immediately read off the well-known isospin-relation [76]

$$\mathcal{A}(B^- \to \bar{K}^0\pi^-) - \sqrt{2}\,\mathcal{A}(B^- \to \bar{K}^-\pi^0) = -\mathcal{A}(\bar{B}^0 \to K^-\pi^+) - \sqrt{2}\,\mathcal{A}(\bar{B}^0 \to \bar{K}^0\pi^0). \quad (6.8)$$

It turns out that $B \to K\pi$ is dominated by the QCD penguin contribution implying $|\mathcal{A}_{1/2}^{\Delta I=0}| \gg |\mathcal{A}_{3/2,\,1/2}^{\Delta I=1}|$. Therefore in a first approximation all the decay modes can be described by a single amplitude $\mathcal{A}_{1/2}^{\Delta I=0}$. This dictates the relative size of the branching fractions to be $2 : 1 : 2 : 1$ (listed in the same order as the amplitudes in (6.7)) and enforces vanishing direct CP violation. Brilliant progress of the B factory experiments allow us today to study deviations from this patterns. In this way we can probe the isospin-violating amplitudes $\mathcal{A}_{3/2,\,1/2}^{\Delta I=1}$.

6.2. Topological parametrisation

In the last section we have used isospin symmetry to parametrise the $B \to K\pi$ amplitudes in terms of the reduced matrix-elements $\mathcal{A}_{1/2}^{\Delta I=0}$, $\mathcal{A}_{3/2,\,1/2}^{\Delta I=1}$. These isospin invariant amplitudes receive contributions from various SM quark diagrams. It is only at the level of these diagrams that the pattern of CP violation can be correctly implemented, i.e. that the amplitudes $\mathcal{A}_{1/2}^{\Delta I=0}$, $\mathcal{A}_{3/2,\,1/2}^{\Delta I=1}$ can be related to their CP conjugated counterparts $\overline{\mathcal{A}}_{1/2}^{\Delta I=0}$, $\overline{\mathcal{A}}_{3/2,\,1/2}^{\Delta I=1}$. This suggests an alternative parametrisation of the amplitudes in terms of the topologies of the underlying quark-level transitions [77]:

$$
\begin{aligned}
\mathcal{A}(B^- \to \bar{K}^0\pi^-) &\simeq P\left(1 - \frac{1}{3}r_{\rm EW}^{C} + \frac{2}{3}r_{\rm EW}^{\rm A}\right), \\
\sqrt{2}\,\mathcal{A}(B^- \to K^-\pi^0) &\simeq P\left(1 + r_{\rm EW} + \frac{2}{3}r_{\rm EW}^{C} + \frac{2}{3}r_{\rm EW}^{\rm A} - (r_{\rm T} + r_{\rm C})e^{-i\gamma}\right), \\
\mathcal{A}(\bar{B}^0 \to K^-\pi^+) &\simeq P\left(1 + \frac{2}{3}r_{\rm EW}^{C} - \frac{1}{3}r_{\rm EW}^{\rm A} - r_{\rm T}e^{-i\gamma}\right), \\
\sqrt{2}\,\mathcal{A}(\bar{B}^0 \to \bar{K}^0\pi^0) &\simeq -P\left(1 - r_{\rm EW} - \frac{1}{3}r_{\rm EW}^{C} - \frac{1}{3}r_{\rm EW}^{\rm A} + r_{\rm C}e^{-i\gamma}\right).
\end{aligned}
\tag{6.9}
$$

This topological parametrisation is illustrated by the corresponding Feynman diagrams for $B^- \to K^-\pi^0$ in Fig. 6.1. In Eq. (6.9) we have factored out the dominant QCD penguin amplitude P and neglected penguin amplitudes suppressed by $|V_{us}^*V_{ub}|/|V_{cs}^*V_{cb}|$. The dependence on the weak CKM phase γ has been made explicit while strong phases are contained in the ratios r_i which fulfill $|r_i| < 1$. These quantities denote corrections from different types of Feynman diagrams: $r_{\rm T}$ and $r_{\rm C}$ stem from colour-allowed and colour-suppressed tree diagrams, $r_{\rm EW}$ and $r_{\rm EW}^{\rm C}$ from colour-allowed and colour-suppressed electroweak penguins, respectively. Weak annihilation via QCD penguin diagrams is absorbed into P whereas weak annihilation via electroweak penguin diagrams is parametrised by $r_{\rm EW}^{\rm A}$ and colour-suppressed tree annihilation is neglected. The r_i are in direct correspondence to the QCDF amplitudes α_i in Eq. (2.12). With our set up of Section 2.2.3 we obtain

$$
\begin{aligned}
r_{\rm T} &= -\left|\frac{\lambda_u}{\lambda_c}\right| \frac{\alpha_1(\pi,K)}{\hat{\alpha}_4^c(\pi,K)} = 0.17^{+0.07}_{-0.06} + 0.03^{+0.03}_{-0.10}\,i\,, \\
r_{\rm C} &= -\left|\frac{\lambda_u}{\lambda_c}\right| \frac{A_{K\pi}}{A_{\pi K}}\frac{\alpha_2(K,\pi)}{\alpha_4^c(\pi,K)} = 0.07^{+0.04}_{-0.06} + (-0.01)^{+0.03}_{-0.05}\,i\,, \\
r_{\rm EW} &= \frac{3}{2}\frac{A_{K\pi}}{A_{\pi K}}\frac{\alpha_{3,\rm EW}^c(K,\pi)}{\hat{\alpha}_4^c(\pi,K)} = 0.13^{+0.05}_{-0.05} + 0.02^{+0.02}_{-0.07}\,i\,, \\
r_{\rm EW}^{\rm C} &= \frac{3}{2}\frac{\alpha_{4,\rm EW}^c(\pi,K)}{\hat{\alpha}_4^c(\pi,K)} = 0.04^{+0.02}_{-0.03} + (-0.01)^{+0.02}_{-0.03}\,i\,, \\
r_{\rm EW}^{\rm A} &= \frac{3}{2}\frac{\beta_{3,\rm EW}^c(\pi,K)}{\hat{\alpha}_4^c(\pi,K)} = 0.007^{+0.002}_{-0.010} + (-0.004)^{+0.011}_{-0.003}\,i\,,
\end{aligned}
\tag{6.10}
$$

where $\lambda_{u,c}$ abbreviate the usual combinations of CKM factors, $A_{K\pi}$ and $A_{\pi K}$ contain form factors and decay constants, and $\hat{\alpha}_4^c = \alpha_4^c + \beta_3$. The results display the typical features of QCDF predictions as discussed in Section 2.2.2, namely small strong phases and large uncertainties of colour-suppressed topologies. The smallness of the r_i reflects the domination of the isospin-conserving QCD penguin and justifies the expansion of physical observables in the r_i. Among the isospin-violating contributions the colour-allowed tree gives the largest corrections followed by the EW penguin which dominates over the colour-suppressed tree. The colour-suppressed EW penguin ratio $r_{\rm EW}^{\rm C}$ and especially the EW penguin annihilation ratio $r_{\rm EW}^{\rm A}$ are quite small and consequently they have been omitted in most analyses of $B \to K\pi$ decays. In particular, the possibility of having NP in the EW penguin annihilation amplitude $r_{\rm EW}^{\rm A}$ has to our knowledge not been considered so far. However, we want to point out that such an approximation is not valid in the analysis of CP asymmetries within the QCDF framework: CP asymmetries are related to the imaginary parts of the r_i which are in QCDF generated at $\mathcal{O}(\alpha_s)$. At this order the colour-suppression of $r_{\rm EW}^{\rm C}$ is not present anymore and the $\Lambda_{\rm QCD}/m_b$ suppressed $r_{\rm EW}^{\rm A}$ can compete as well. Therefore we keep $r_{\rm EW}^{\rm C}$ and $r_{\rm EW}^{\rm A}$ in order to allow for the possibility to have a large NP contribution to these amplitudes.

By comparing Eqs. (6.7) and (6.9) one can express the isospin amplitudes through the topological parameters. This results in

$$\begin{aligned}
\mathcal{A}_{1/2}^{\Delta I=0} &= P\left(1 + \frac{1}{6}\,r_{\rm EW}^{\rm C} + \frac{1}{6}r_{\rm EW}^{\rm A} - \frac{1}{2}\,r_{\rm T}\,e^{-i\gamma}\right), \\
\mathcal{A}_{3/2}^{\Delta I=1} &= P\left(\frac{1}{3}\,r_{\rm EW} + \frac{1}{3}\,r_{\rm EW}^{\rm C} - \left(\frac{1}{3}\,r_{\rm T} + \frac{1}{3}\,r_{\rm C}\right)e^{-i\gamma}\right), \\
\mathcal{A}_{1/2}^{\Delta I=1} &= P\left(\frac{1}{3}\,r_{\rm EW} - \frac{1}{6}\,r_{\rm EW}^{\rm C} + \frac{1}{2}r_{\rm EW}^{\rm A} + \left(\frac{1}{6}\,r_{\rm T} - \frac{1}{3}\,r_{\rm C}\right)e^{-i\gamma}\right).
\end{aligned} \tag{6.11}$$

Let us for the moment neglect the small $r_{\rm EW}^{\rm A}$ and focus on the two $\Delta I = 1$ amplitudes $\mathcal{A}_{3/2}^{\Delta I=1}$ and $\mathcal{A}_{1/2}^{\Delta I=1}$. Obviously they can be written as linear combinations of

$$r_{\rm EW} - r_{\rm C}\,e^{-i\gamma} \qquad \text{and} \qquad r_{\rm EW}^{\rm C} - r_{\rm T}\,e^{-i\gamma}. \tag{6.12}$$

This implies that colour-allowed EW penguins and colour-suppressed trees, on the one hand, as well as colour-suppressed EW penguins and colour-allowed trees, on the other hand, form pairs of contributions which are inextricably linked to each other as far as purely isospin-violating observables are concerned. If we moreover allow for arbitrary magnitudes, weak and strong phases beyond the SM for all the topologies, we find the topological parametrisation to contain some redundancy: Whereas the isospin amplitudes provide only 8 free parameters, the complex $\mathcal{A}_{3/2,1/2}^{\Delta I=1}$ and $\overline{\mathcal{A}}_{3/2,1/2}^{\Delta I=1}$, the topological parametrisation introduces 12 free parameters, the magnitudes, weak and strong phases of the four different topologies. Therefore physical effects found in any experiment cannot unambiguously be attributed to one or the other partner of the topology pairs in (6.12). For the pair consisting of $r_{\rm T}$ and $r_{\rm EW}^{\rm C}$ this ambiguity is resolved by $\mathcal{A}_{1/2}^{\Delta I=0}$ but $r_{\rm EW}$ and $r_{\rm C}$ stay inseparable.

In the next section we will discuss hints on NP in the EW penguin sector. Such contributions can be included into all formulae given in this section by the replacements

$$r_{\rm EW} \to r_{\rm EW} + \tilde{r}_{\rm EW} e^{-i\delta}, \qquad r_{\rm EW}^{\rm C} \to r_{\rm EW}^{\rm C} + \tilde{r}_{\rm EW}^{\rm C} e^{-i\delta}, \qquad r_{\rm EW}^{\rm A} \to r_{\rm EW}^{\rm A} + \tilde{r}_{\rm EW}^{\rm A} e^{-i\delta}, \quad (6.13)$$

where the $\tilde{r}_{\rm EW}^{i}$ are complex parameters containing strong phases and δ is a new weak phase. Assuming that there is no NP in tree topologies and that the SM amplitudes can be calculated from theory, the $\tilde{r}_{\rm EW}^{i}$ and δ are then accessible from experiment. For example the three parameters $\mathrm{Re}(\tilde{r}_{\rm EW})$, $\mathrm{Im}(\tilde{r}_{\rm EW})$ and δ can be determined from the four physical combinations

$$\begin{aligned}
&\mathrm{Re}(\tilde{r}_{\rm EW})\cos\delta \; - \; \mathrm{Re}(r_{\rm C})\cos\gamma \; + \; \mathrm{Re}(r_{\rm EW})\,,\\
&\mathrm{Im}(\tilde{r}_{\rm EW})\sin\delta \; - \; \mathrm{Im}(r_{\rm C})\sin\gamma\,,\\
&\mathrm{Im}(\tilde{r}_{\rm EW})\cos\delta \; - \; \mathrm{Im}(r_{\rm C})\cos\gamma \; + \; \mathrm{Im}(r_{\rm EW})\,,\\
&\mathrm{Re}(\tilde{r}_{\rm EW})\sin\delta \; - \; \mathrm{Re}(r_{\rm C})\sin\gamma\,. \qquad (6.14)
\end{aligned}$$

However, one should keep in mind that hadronic uncertainties can mimic (or hide) such a NP signal and this concerns also hadronic uncertainties from the SM trees since they are linked to the EW penguins as it has been discussed above and as it is manifest in (6.14). Therefore, probing a colour-allowed NP contribution $\tilde{r}_{\rm EW}$ is challenged by the large hadronic uncertainties in the QCDF prediction for $r_{\rm C}$.

6.3. Current status of isospin violation in $B \to K\pi$

The dominance of the QCD penguin topology in the $B \to K\pi$ amplitudes establishes an approximate isospin symmetry of the amplitudes which manifests itself in a number of sum rules linking the branching fractions [78] or direct CP asymmetries [79] of the various decay modes. These sum rules can be exploited to construct measurable quantities which are sensitive to isospin violation. In this section we will define corresponding observables and derive approximate formulae for them in terms of the topological parameters r_i. We will calculate their SM values using QCDF, compare the results with the current experimental data and discuss the impact of potential NP in the EW penguin sector. In subsequent chapters we will use these observables then in a twofold way: On the one hand, we will deduce from them simple 2σ constraints on the free parameters of various NP scenarios. On the other hand, we use independent subsets of these observables to perform a χ^2 fit which allows to extract the preferred values of the NP parameters. It should be stressed that the approximate formulae given for the observables in this section serve only to identify their sensitivity to different topological contributions whereas we use the exact expressions for our numerical calculations.

Observable	Theory	Experiment
Br($\bar{B}^0 \to \bar{K}^0\pi^0$) $\times 10^6$	$5.8^{+5.7}_{-3.6}$	$9.5^{+0.5}_{-0.5}$
Br($\bar{B}^0 \to K^-\pi^+$) $\times 10^6$	$14.0^{+12.1}_{-7.8}$	$19.4^{+0.6}_{-0.6}$
Br($B^- \to K^-\pi^0$) $\times 10^6$	$9.6^{+7.3}_{-4.9}$	$12.9^{+0.6}_{-0.6}$
Br($B^- \to \bar{K}^0\pi^-$) $\times 10^6$	$15.7^{+13.7}_{-8.9}$	$23.1^{+1.0}_{-1.0}$
$R_c^B(K\pi)$	$1.22^{+0.17}_{-0.15}$	$1.12^{+0.07}_{-0.07}$
$R_n^B(K\pi)$	$1.22^{+0.18}_{-0.16}$	$1.02^{+0.06}_{-0.06}$
$R_c^K(K\pi)$	$1.27^{+0.12}_{-0.11}$	$1.24^{+0.07}_{-0.07}$
$R_n^K(K\pi)$	$1.27^{+0.15}_{-0.13}$	$1.13^{+0.08}_{-0.07}$
$R_c^\pi(K\pi)$	$1.04^{+0.10}_{-0.08}$	$1.11^{+0.06}_{-0.06}$
$R_n^\pi(K\pi)$	$1.55^{+0.38}_{-0.31}$	$1.26^{+0.09}_{-0.09}$
$R(K\pi)$	$1.02^{+0.02}_{-0.02}$	$1.05^{+0.05}_{-0.05}$
$A_{\text{CP}}(\bar{B}^0 \to \bar{K}^0\pi^0)$	$-0.003^{+0.057}_{-0.108}$	$-0.01^{+0.10}_{-0.10}$
$A_{\text{CP}}(\bar{B}^0 \to K^-\pi^+)$	$-0.047^{+0.187}_{-0.047}$	$-0.098^{+0.012}_{-0.011}$
$A_{\text{CP}}(B^- \to K^-\pi^0)$	$-0.028^{+0.221}_{-0.059}$	$0.050^{+0.025}_{-0.025}$
$A_{\text{CP}}(B^- \to \bar{K}^0\pi^-)$	$0.003^{+0.012}_{-0.003}$	$0.009^{+0.025}_{-0.025}$
$\Delta A_{\text{CP}} = \Delta A^-_{\text{CP}}$	$0.019^{+0.058}_{-0.048}$	$0.148^{+0.027}_{-0.028}$
ΔA^0_{CP}	$0.006^{+0.118}_{-0.057}$	$0.019^{+0.103}_{-0.103}$
$S_{\text{CP}}(\bar{B}^0 \to \bar{K}^0\pi^0)$	$0.80^{+0.06}_{-0.08}$	$0.57^{+0.17}_{-0.17}$

Table 6.1: Theoretical versus experimental results for the $\bar{B} \to \bar{K}\pi$ decays. The experimental data is taken from [42].

6.3.1. Direct CP asymmetries

Non-vanishing direct CP asymmetries are caused by the interference of parts of the decay amplitude which have both, different weak and different strong phases. Therefore direct CP asymmetries in $B \to K\pi$ cannot be generated by the single QCD penguin amplitude and are automatically sensitive to subleading contributions. Neglecting terms quadratic in the r_i we find for the $B \to K\pi$ decay modes

$$\begin{aligned}
A_{\text{CP}}(B^- \to \bar{K}^0\pi^-) &\simeq 0, \\
A_{\text{CP}}(B^- \to K^-\pi^0) &\simeq -2\,\text{Im}\,(r_{\text{T}} + r_{\text{C}}) \sin\gamma, \\
A_{\text{CP}}(\bar{B}^0 \to K^-\pi^+) &\simeq -2\,\text{Im}(r_{\text{T}}) \sin\gamma, \\
A_{\text{CP}}(\bar{B}^0 \to \bar{K}^0\pi^0) &\simeq 2\,\text{Im}\,(r_{\text{C}}) \sin\gamma.
\end{aligned} \tag{6.15}$$

If one assumes r_T to dominate over r_C, the asymmetries $A_{CP}(\bar{B}^0 \to K^-\pi^+)$ and $A_{CP}(B^- \to K^-\pi^0)$ are expected to have the same sign and to be approximately equal. This expectation would be reflected in a nearly vanishing

$$\Delta A_{CP} \equiv \Delta A_{CP}^- \equiv A_{CP}(B^- \to K^-\pi^0) - A_{CP}(\bar{B} \to K^-\pi^+) \simeq -2\,\mathrm{Im}\,(r_C)\sin\gamma. \quad (6.16)$$

Current experimental data, however, show different signs for the two asymmetries (see table 6.1) and yields

$$\Delta A_{CP} \overset{\text{exp.}}{=} (14.8 \pm 2.8)\%. \quad (6.17)$$

The only possible explanation for a large ΔA_{CP} in the SM would be a large imaginary part of r_C. QCDF predicts only a small $\mathrm{Im}(r_C)$, even when all the theory errors are included, and we obtain

$$\Delta A_{CP} \overset{\text{SM}}{=} 1.9^{+5.8}_{-4.8}\,\%. \quad (6.18)$$

Adopting a frequentist approach where we give not preference to any theory value within the error interval (but consider the true value to lie definitely within the error interval), we find a $\sim 2.5\,\sigma$ discrepancy between theory and experiment.

Now let us see how the situation changes if we include NP in the EW penguin sector according to (6.13). Provided the phase δ is not zero, the new contributions enters the CP asymmetries as

$$\begin{aligned}
A_{CP}(B^- \to \bar{K}^0\pi^-) &\simeq -2\,\mathrm{Im}(\frac{1}{3}\tilde{r}_{EW}^C + \frac{2}{3}\tilde{r}_{EW}^A)\sin\delta,\\
A_{CP}(B^- \to K^-\pi^0) &\simeq -2\,\mathrm{Im}\,(r_T + r_C)\sin\gamma + 2\,\mathrm{Im}\left(\tilde{r}_{EW} + \frac{2}{3}\tilde{r}_{EW}^C + \frac{2}{3}\tilde{r}_{EW}^A\right)\sin\delta,\\
A_{CP}(\bar{B}^0 \to K^-\pi^+) &\simeq -2\,\mathrm{Im}(r_T)\sin\gamma + 2\,\mathrm{Im}(\frac{2}{3}\tilde{r}_{EW}^C - \frac{1}{3}\tilde{r}_{EW}^A)\sin\delta,\\
A_{CP}(\bar{B}^0 \to \bar{K}^0\pi^0) &\simeq 2\,\mathrm{Im}\,(r_C)\sin\gamma - 2\,\mathrm{Im}\left(\tilde{r}_{EW} + \frac{1}{3}\tilde{r}_{EW}^C + \frac{1}{3}\tilde{r}_{EW}^A\right)\sin\delta
\end{aligned} \quad (6.19)$$

and ΔA_{CP} as

$$\Delta A_{CP} \simeq -2\,\mathrm{Im}\,(r_C)\sin\gamma + 2\,\mathrm{Im}\,(\tilde{r}_{EW} + \tilde{r}_{EW}^A)\sin\delta. \quad (6.20)$$

With the additional contributions from $\tilde{r}_{EW}$ and $\tilde{r}_{EW}^A$ the difference ΔA_{CP} can turn out much larger than in the SM. We will see that the observed discrepancy in ΔA_{CP} can be solved by $\tilde{r}_{EW}$ as well as by $\tilde{r}_{EW}^A$. To this end one needs NP in the EW penguin sector of the order of the SM Wilson coefficient C_9^{SM}

Apart from ΔA_{CP} we can construct a second difference

$$\begin{aligned}
\Delta A_{CP}^0 &\equiv A_{CP}(B^- \to \bar{K}^0\pi^-) - A_{CP}(\bar{B} \to \bar{K}^0\pi^0)\\
&\simeq -2\,\mathrm{Im}\,(r_C)\sin\gamma + 2\,\mathrm{Im}\,(\tilde{r}_{EW} + \tilde{r}_{EW}^A)\sin\delta\,,
\end{aligned} \quad (6.21)$$

which in principle could also be used to probe $\tilde{r}_{\rm EW}$ and $\tilde{r}_{\rm EW}^{\rm A}$. Moreover, the difference of $\Delta A_{\rm CP}^{-}$ and $\Delta A_{\rm CP}^{0}$ would even be sensitive to terms quadratic in the r_i [79]. Unfortunately, data on $A_{\rm CP}(B^- \to \bar{K}^0\pi^-)$ and especially on $A_{\rm CP}(\bar{B}^0 \to \bar{K}^0\pi^0)$ are not good enough yet to gain any information from these observables.

6.3.2. Branching fractions

We already noticed in Section 6.1 that ratios of any two different decay rates measure isospin violation. A further advantage of considering such ratios is that form factors and decay constants cancel and this decreases the uncertainties of the QCDF predictions. Using the parametrisation (6.9) and neglecting terms which are quadratic in the r_i as well as the annihilation contribution $r_{\rm EW}^{\rm A}$ which has only a small real part, the six different ratios read [31]

$$
\begin{aligned}
R_c^B &\equiv 2\,\frac{\overline{\rm Br}(B^- \to K^-\pi^0)}{\overline{\rm Br}(B^- \to \bar{K}^0\pi^-)} \simeq 1 + 2\,{\rm Re}(r_{\rm EW} + r_{\rm EW}^{\rm C}) - 2\,{\rm Re}(r_{\rm T} + r_{\rm C})\cos\gamma\,, \\
R_n^B &\equiv \frac{1}{2}\,\frac{\overline{\rm Br}(\bar{B}^0 \to K^-\pi^+)}{\overline{\rm Br}(\bar{B}^0 \to \bar{K}^0\pi^0)} \simeq 1 + 2\,{\rm Re}(r_{\rm EW} + r_{\rm EW}^{\rm C}) - 2\,{\rm Re}(r_{\rm T} + r_{\rm C})\cos\gamma\,, \\
R_c^K &\equiv 2\,\frac{\tau_0}{\tau_-}\,\frac{\overline{\rm Br}(B^- \to K^-\pi^0)}{\overline{\rm Br}(\bar{B}^0 \to K^-\pi^+)} \simeq 1 + 2\,{\rm Re}(r_{\rm EW}) - 2\,{\rm Re}(r_{\rm C})\cos\gamma\,, \\
R_n^K &\equiv \frac{1}{2}\,\frac{\tau_0}{\tau_-}\,\frac{\overline{\rm Br}(B^- \to \bar{K}^0\pi^-)}{\overline{\rm Br}(\bar{B}^0 \to \bar{K}^0\pi^0)} \simeq 1 + 2\,{\rm Re}(r_{\rm EW}) - 2\,{\rm Re}(r_{\rm C})\cos\gamma\,, \\
R_c^\pi &\equiv \frac{\tau_0}{\tau_-}\,\frac{\overline{\rm Br}(B^- \to \bar{K}^0\pi^-)}{\overline{\rm Br}(\bar{B}^0 \to K^-\pi^+)} \simeq 1 + 2\,{\rm Re}(r_{\rm T})\cos\gamma - 2\,{\rm Re}(r_{\rm EW}^{\rm C}), \\
R_n^\pi &\equiv \frac{\tau_0}{\tau_-}\,\frac{\overline{\rm Br}(B^- \to K^-\pi^0)}{\overline{\rm Br}(\bar{B}^0 \to \bar{K}^0\pi^0)} \simeq 1 - 2\,{\rm Re}(r_{\rm T} + 2r_{\rm C})\cos\gamma + 2\,{\rm Re}(2r_{\rm EW} + r_{\rm EW}^{\rm C})\,. \qquad (6.22)
\end{aligned}
$$

Here $\overline{\rm Br}$ denotes CP-averaged branching ratios, τ_0 and τ_- are the life times of the neutral and charged B mesons, respectively. If one finds deviations of these ratios from $R_{c,n}^{B,K,\pi} = 1$ which cannot be explained by the SM values of the r_i, this can be interpreted as a hint on NP in EW penguins entering the ratios $R_{c,n}^{B,K,\pi}$ through

$$
\begin{aligned}
{\rm Re}(r_{\rm EW}) &\to {\rm Re}(r_{\rm EW}) + {\rm Re}(\tilde{r}_{\rm EW})\cos\delta, \\
{\rm Re}(r_{\rm EW}^{\rm C}) &\to {\rm Re}(r_{\rm EW}^{\rm C}) + {\rm Re}(\tilde{r}_{\rm EW}^{\rm C})\cos\delta. \qquad (6.23)
\end{aligned}
$$

Indeed a discrepancy in the early data displaying $R_c \equiv R_c^B > 1$ and $R_n \equiv R_n^B < 1$ raised the formulation of a "$B \to K\pi$ puzzle" in the first place and we see from (6.22) that even terms quadratic in the r_i are needed to account for this pattern. In the meantime, the measurements fluctuated towards the SM values and we find our QCDF results for the $R_{c,n}^{B,K,\pi}$ to be in good agreement with the current experimental data (see table 6.1). However, if there exist NP contributions $\tilde{r}_{\rm EW}$ and $\tilde{r}_{\rm EW}^{\rm C}$, as suggested by $\Delta A_{\rm CP}$, they will be constrained from the $R_{c,n}^{B,K,\pi}$. Note that the $R_{c,n}^{B,K,\pi}$ involve different combinations of $\tilde{r}_{\rm EW}$ and $\tilde{r}_{\rm EW}^{\rm C}$ and thus they are sensitive to different linear combinations of the electroweak penguin coefficients $C_7^{(\prime)}, \ldots, C_{10}^{(\prime)}$. Therefore, it depends on the specific NP scenario in consideration which of the $R_{c,n}^{B,K,\pi}$ give the best constraints.

Beyond being responsible for the universality of the QCD penguin contribution, isospin relations account for the approximate equation

$$\Gamma(B^- \to \pi^- \bar{K}^0) - 2\,\Gamma(B^- \to \pi^0 K^-) \;\approx\; 2\,\Gamma(\bar{B}^0 \to \pi^0 \bar{K}^0) - \Gamma(\bar{B}^0 \to \pi^+ K^-) \qquad (6.24)$$

known as Lipkin sum rule. In the strict isospin limit both sides of this equation vanish identically and this is reflected in the fact that $R_{c,n}^{B}$ in Eq. (6.22) is equal to one apart from isospin-violating terms of order $\mathcal{O}(r_i)$. These terms which are linear in the r_i are generated by the interference of the isospin-violating parts of the amplitude with the QCD penguin part. The special property of Eq. (6.24) is now that these interference terms on the left- and righthand side of the approximate equation cancel each other. For this reason Eq. (6.24) can be used to construct a purely isospin-violating observable, namely

$$R \equiv 2\,\frac{\tau_-\,\overline{\rm Br}(\bar{B}^0 \to \pi^0 \bar{K}^0) \,+\, \tau_0\,\overline{\rm Br}(B^- \to \pi^0 K^-)}{\tau_-\,\overline{\rm Br}(\bar{B}^0 \to \pi^+ K^-) \,+\, \tau_0\,\overline{\rm Br}(B^- \to \pi^- \bar{K}^0)} \;=\; 1 \,+\, \mathcal{O}(r_i^2)\,. \qquad (6.25)$$

Also for this observable we find agreement between the experimental value and our QCDF prediction. It can be used as a further constraint on $\tilde{r}_{\rm EW}$ and $\tilde{r}_{\rm EW}^{\rm C}$.

6.3.3. Mixing-induced CP violation

Since $K_s\pi^0$ is a CP-eigenstate into which both the B^0 and the $\bar{B}^0$ meson can decay, we have mixing-induced CP violation in this decay channel. For details on the mechanism of mixing-induced CP violation and the definition of the corresponding observable $S_{K\pi}$ we refer to Section 5.3. Although $S_{K\pi}$ is not sensitive to isospin-violation in particular, it will be affected by a solution of the "$\Delta A_{\rm CP}$-puzzle" via a NP contribution $\tilde{r}_{\rm EW}$. The reason is that $\tilde{r}_{\rm EW}$ has to come with a large new weak phase δ in order to have substantial impact on $\Delta A_{\rm CP}$. With our parametrisation we find for the modified $S_{K\pi}$:

$$S_{K\pi} \simeq \sin 2\beta + 2{\rm Re}\,(r_{\rm C}) \cos 2\beta \sin\gamma - 2{\rm Re}(\tilde{r}_{\rm EW} + \tilde{r}_{\rm EW}^{C}) \cos 2\beta \sin\delta. \qquad (6.26)$$

Here we have neglected again terms quadratic in the r_i and a term proportional to the small ${\rm Re}(\tilde{r}_{\rm EW}^{\rm A})$. The error bands of experimental and theoretical values for $S_{K\pi}$ rarely overlap (see table 6.1). This allows for some freedom in $\tilde{r}_{\rm EW}$ which is needed to explain $\Delta A_{\rm CP}$ through this parameter.

To summarise: We have studied various observables which are sensitive to the EW penguin contribution $r_{\rm EW}$ which enters always together with r_c in one of the four combinations anticipated in (6.14). The only observable seriously pointing to a new contribution $\tilde{r}_{\rm EW}$ so far is $\Delta A_{\rm CP}$. Hence in order to clarify the situation one should study further decays.

6.4. The decays $B \to K\rho$, $K^*\pi$, $K^*\rho$

The decays $B \to K\rho, K^*\pi, K^*\rho$ are simply the PV, VP, VV counterparts of $B \to K\pi$, with which they share the flavour structure. Since our analysis of $B \to K\pi$ did not give a clear picture of the EW penguin sector, it could be enlightening to complement it with a study of the corresponding PV, VP, VV modes. These decay channels are even more sensitive to isospin violation because the dominating QCD penguin amplitude is smaller resulting in larger r_i ratios. To exemplify this we give here the r_i for $B \to K\rho$:

$$\begin{aligned} r_{\rm T}(\rho\bar{K}) &= -0.34^{+0.22}_{-0.49} - 0.28^{+0.69}_{-0.22}\, i\,, \\ r_{\rm C}(\rho\bar{K}) &= -0.20^{+0.17}_{-0.21} - 0.09^{+0.36}_{-0.11}\, i\,, \\ r_{\rm EW}(\rho\bar{K}) &= -0.33^{+0.21}_{-0.47} - 0.27^{+0.68}_{-0.21}\, i\,, \\ r^{C}_{\rm EW}(\rho\bar{K}) &= -0.11^{+0.09}_{-0.11} - 0.05^{+0.19}_{-0.06}\, i\,. \end{aligned} \tag{6.27}$$

However, experimental data of these decays is not yet precise enough to draw any significant conclusions. Nevertheless these decays can give constraints which are complementary to those from $B \to K\pi$: Because of their different spin structure, they probe other linear combinations of the EW penguin coefficients $C_7^{(\prime)}, ..., C_{10}^{(\prime)}$ (see Eqs. (2.12) and (2.12)). In our analyses we include them therefore as $2\,\sigma$ constraints but because of the insufficient precision of data we decided not to incorporate them in our χ^2 fits. Tables containing our QCDF predictions and the experimental values for the $B \to K\rho, K^*\pi, K^*\rho$ observables can be found in appendix A.5.

7. THE PURELY ISOSPIN VIOLATING DECAYS $B_s \to \phi\pi, \phi\rho$

In the previous chapter we suggested that, in order to find out whether the discrepancy in the $B \to K\pi$ observable $\Delta A_{\rm CP}$ is provoked by NP in electroweak penguins, one should study also other decays which are sensitive to this sector. This is not an easy task, since EW penguin contributions are usually overshadowed by the larger QCD penguins. This problem can be avoided if one succeeds in probing exclusively the $\Delta I = 1$ part of the Hamiltonian which is orthogonal to the QCD penguin operators. To achieve this for $B \to K\pi$ we had to single out the $\Delta I = 1$-part of (6.5) by combining different isospin-related decay modes. Our proposal is now to consider decays to which QCD penguins do not contribute at all, i.e. pure $\Delta I = 1$ decays, where no such procedure is needed.

There are no two-body decays of the B_d or $B^\pm$ meson with this property. In these cases the final state would have to be a pure $|3/2, \pm 1/2\rangle$ isospin state which cannot be constructed out of two mesons. The B_s meson, on the other hand, is an isosinglet and it can decay as

$$0 \overset{\Delta I=1}{\longrightarrow} 0 \otimes 1 = 1 \,. \tag{7.1}$$

The final state must consist of an isospin triplet, i.e. π^0 or ρ^0, and an isosinglet, i.e. a meson with the $s\bar{s}$ flavour structure. In order to avoid complications stemming from $\eta - \eta'$-mixing, we restrict ourselves to the vector-meson ϕ which is to a good approximation a pure $s\bar{s}$ state. This leaves us with the two $\Delta I = 1$-channels $B_s \to \phi\pi^0$ and $B_s \to \phi\rho^0$.

These decays have a very simple topological structure which is depicted in Fig. 7.1. They can be parametrised as

$$\sqrt{2}\, A(\bar{B}_s \to \phi\pi/\rho) \,=\, P_{\rm EW}^{\pi/\rho} \left(1 \,-\, r_{\rm C}^{\pi/\rho}\, e^{-i\gamma}\right) \tag{7.2}$$

where we have factored out the EW penguin amplitude $P_{\rm EW}^{\pi/\rho}$ anticipating its dominance over the colour-suppressed tree represented by the tree-to-penguin ratio $r_{\rm C}^{\pi/\rho}$. In terms of QCDF amplitudes this ratio reads

$$r_{\rm C}^{\pi/\rho} = -\frac{2}{3} \left| \frac{\lambda_u^{(s)}}{\lambda_c^{(s)}} \right| \frac{\alpha_2(\phi, \pi/\rho)}{\alpha_{3,\rm EW}(\phi, \pi/\rho)} \,. \tag{7.3}$$

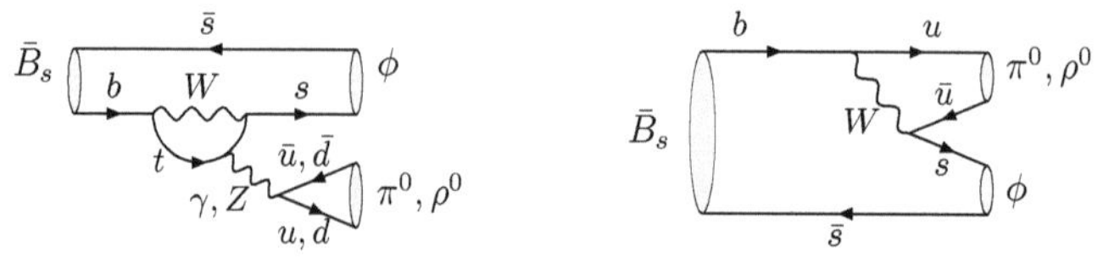

Figure 7.1: EW penguin (r_{EW}) and colour-suppressed tree (r_{C}) contributions to $B_s \to \phi\pi, \phi\rho$

Its numerical value is given by

$$\begin{aligned} r_{\text{C}}^{\pi} &= 0.41^{+0.37}_{-0.41} - 0.13^{+0.30}_{-0.30}\, i\,, \\ r_{\text{C}}^{\rho,0} &= 0.39^{+0.35}_{-0.39} - 0.13^{+0.28}_{-0.29}\, i\,, \\ r_{\text{C}}^{\rho,-} &= 0.21^{+0.49}_{-0.46} + 0.15^{+0.45}_{-0.45}\, i\,, \end{aligned} \tag{7.4}$$

for the isotriplet meson being π^0, a longitudinally polarised ρ^0 and a negatively polarised ρ^0, respectively. Choosing a phase convention such that $P_{\text{EW}}^{\pi/\rho}$ is real, we further find

$$P_{\text{EW}}^{\pi} = 6.45^{+1.87}_{-0.54} \cdot 10^{-9}, \qquad P_{\text{EW}}^{\rho,0} = 9.95^{+2.83}_{-0.79} \cdot 10^{-9}, \qquad P_{\text{EW}}^{\rho,-} = 4.27^{+1.34}_{-0.81} \cdot 10^{-9}. \tag{7.5}$$

The positive helicity amplitude has been neglected according to its $\Lambda_{\text{QCD}}^2/m_b^2$ suppression (see discussion in Section 2.2.2). Weak annihilation contributions have been omitted from Eq. (7.2). They are expected to be small, even beyond the usual Λ_{QCD}/m_b suppression, because a colour singlet $\bar{s}s$ pair forming the ϕ meson must be created from gluons in this case. The corresponding diagrams are suppressed by the OZI rule [80] and by higher powers of α_s since three gluons are needed to match the quantum numbers of the ϕ.

The fact that $|r_{\text{C}}^{\pi/\rho}| < 1$ shows that these decays are indeed dominated by the EW penguin topology and suggests them as the ideal candidates (golden channels) to test the hypothesis of NP in this sector. A new contribution to the $B \to K\pi$ amplitudes of the form (6.13) would enter also the $B_s \to \phi\pi, \phi\rho$ amplitude (7.2) modifying it as

$$\sqrt{2}\, A(\bar{B}_s \to \phi\pi/\rho) = P_{\text{EW}}^{\pi/\rho} \left(1 - r_{\text{C}}^{\pi/\rho}\, e^{-i\gamma} + \tilde{r}_{\text{EW}}^{\pi/\rho}\, e^{-i\delta}\right) \tag{7.6}$$

where $\tilde{r}_{\text{EW}}^{\pi/\rho}$ contains a strong phase and δ is the weak phase introduced in (6.13). If we assume the new contribution to be of the order of the SM EW penguin, as a solution of the "ΔA_{CP}-puzzle" requires, we have $|\tilde{r}_{\text{EW}}^{\pi/\rho}| = \mathcal{O}(1)$ and expect an order of magnitude enhancement of the $B_s \to \phi\pi, \phi\rho$ branching fractions.

Exactly as in the $B \to K\pi$ amplitudes the EW penguin contribution $\tilde{r}_{\text{EW}}^{\pi/\rho}$ is again accompanied by the colour-suppressed tree $r_{\rm C}^{\pi/\rho}$ such that uncertainties in the SM prediction of the latter reduce the sensitivity to the former. Let us perform a rough estimation of this $r_{\rm C}^{\pi/\rho}$ pollution: The branching fractions of $B_s \to \phi\pi, \phi\rho$ are dominated by the real parts $\mathrm{Re}(r_{\rm C}^{\pi/\rho})$ and $\mathrm{Re}(\tilde{r}_{\rm EW}^{\pi/\rho})$ since the imaginary parts are small and enter only quadratic. Assuming $\tilde{r}_{\rm EW}^{\pi/\rho} = 1$, which means that the NP contribution is as large as the SM EW penguin and has the same strong phase, and assuming in addition $\delta = \gamma$ for simplicity, the relative importance of the colour-suppressed tree compared to the new EW penguin is determined by $|\mathrm{Re}(r_{\rm C}^{\pi/\rho})|$. From the numerical values

$$|\mathrm{Re}(r_{\rm C}^{\pi})| = 0.41^{+0.37}_{-0.41}, \qquad \left|\mathrm{Re}(r_{\rm C}^{\rho,0})\right| = 0.39^{+0.35}_{-0.39}, \qquad \left|\mathrm{Re}(r_{\rm C}^{\rho,-})\right| = 0.21^{+0.49}_{-0.21}, \tag{7.7}$$

we see that, in order to assign to $r_{\rm C}^{\pi/\rho}$ an effect caused by $\tilde{r}_{\rm EW}^{\pi/\rho} = 1$, one would have to overshoot its default value by about three times the upper error estimate. As this rough estimation has demonstrated, it is unlikely that an order of magnitude enhancement in the branching ratios $B_s \to \phi\pi, \phi\rho$ originates in hadronic uncertainties of the colour-suppressed tree amplitude. Let us compare this situation with $B \to K\pi$: The considered scenario with $\tilde{r}_{\rm EW}^{\pi/\rho} = 1$ corresponds to $\tilde{r}_{\rm EW} = r_{\rm EW}$ for the $B \to K\pi$ parameters in Eqs. (6.9) and (6.13) and the relevance of the colour-suppressed tree is then determined by

$$\left|\frac{\mathrm{Re}(r_{\rm C})}{\mathrm{Re}(r_{\rm EW})}\right| = 0.54^{+0.32}_{-0.40}, \qquad \left|\frac{\mathrm{Im}(r_{\rm C})}{\mathrm{Im}(r_{\rm EW})}\right| = 0.33^{+2.47}_{-0.33} \tag{7.8}$$

where the ratio of the real parts is relevant for the observables $R_{c,n}^{B,K,\pi}$ and $S_{K\pi}$ whereas the ratio of the imaginary parts is relevant for $\Delta A_{\rm CP}$. For the observables $R_{c,n}^{B,K,\pi}$ and $S_{K\pi}$ we find moderate $r_{\rm C}$ pollution similar to $\mathrm{Br}(B_s \to \phi\pi, \phi\rho)$. For $\Delta A_{\rm CP}$, however, the situation is not that clear due to the large uncertainties of the imaginary parts.

We conclude this section quoting our QCDF results for the SM values of the $B_s \to \phi\pi, \phi\rho$ observables. For the CP-averaged branching fractions we obtain

$$\mathrm{Br}(\bar{B}_s \to \phi\pi^0) = 1.6^{+1.1}_{-0.3} \cdot 10^{-7}, \qquad \mathrm{Br}(\bar{B}_s \to \phi\rho^0) = 4.4^{+2.7}_{-0.7} \cdot 10^{-7}. \tag{7.9}$$

Due to the strong suppression of annihilation topologies, one ends up with the same results (up to the stated accuracy) when using the approximate formula (7.2) for the amplitudes, with the numbers for $P_{\rm EW}^{\pi/\rho}$ and $r_{\rm C}^{\pi/\rho}$ given in Eqs. (7.4) and (7.5). Adding the errors of $P_{\rm EW}^{\pi/\rho}$ and $r_{\rm C}^{\pi/\rho}$ in quadrature gives also a reasonable result for the uncertainty of the branching ratios since correlations between the EW penguin and the colour-suppressed tree amplitude drop out in the ratio $r_{\rm C}^{\pi/\rho}$. Because of smallness of the branching ratios which is due to the absence of QCD penguin and colour-allowed tree contributions, these decays have not been observed yet. However, they are in reach of LHCb or a potential SuperB factory, especially if they are enhanced by NP. The branching ratio $\mathrm{Br}(\bar{B}_s \to \phi\rho^0)$ is dominated by decays into longitudinally polarised vector contributing

$$\mathrm{Br_L}(\bar{B}_s \to \phi\rho^0) = 3.7^{+2.5}_{-0.7} \cdot 10^{-7} . \tag{7.10}$$

This corresponds to a longitudinal polarisation fraction of

$$f_\mathrm{L} = 0.84^{+0.08}_{-0.11} . \tag{7.11}$$

One of the main sources of uncertainty in the QCDF predictions for the branching fractions is the form factor $A_0^{B_s\to\phi}$. It can in principle be eliminated by considering the ratios

$$\frac{\mathrm{Br}(\bar{B}_s \to \phi\rho^0)}{\mathrm{Br}(\bar{B}_s \to \phi\pi^0)} = 2.83^{+0.35}_{-0.23}, \qquad \frac{\mathrm{Br}_L(\bar{B}_s \to \phi\rho^0)}{\mathrm{Br}(\bar{B}_s \to \phi\pi^0)} = 2.38^{+0.10}_{-0.08} . \tag{7.12}$$

Finally, we have calculated the direct CP asymmetries obtaining

$$A_\mathrm{CP}(\bar{B}_s \to \phi\pi^0) = 0.27^{+0.50}_{-0.62}, \qquad A_\mathrm{CP}(\bar{B}_s \to \phi\rho^0) = 0.19^{+0.53}_{-0.61} . \tag{7.13}$$

8. MODEL INDEPENDENT ANALYSIS

In the previous chapter we proposed to test the hypothesis of NP in the EW penguin sector, as suggested by the discrepancy in the $B \to K\pi$ observable ΔA_{CP}, by a measurement of the decays $B_s \to \phi\pi, \phi\rho$. In this chapter we support our proposal by a quantitative analysis pursuing the following strategy: We parametrise NP in EW penguins in a model independent way by adding corresponding terms to the Wilson coefficients $C_7^{(\prime)}, ..., C_{10}^{(\prime)}$. By performing a χ^2-fit we determine then the NP parameters in such a way that they describe well the $B \to K\pi$ data, i.e. in particular solve the ΔA_{CP} discrepancy. Further hadronic decays like $B \to K\rho, K^*\pi, K^*\rho$ are used to impose additional constraints at the $2\,\sigma$ level. With respect to the resulting fit we study then the decays $B_s \to \phi\pi, \phi\rho$ and quantify a potential enhancement of their branching fractions. Note that such an exhaustive analysis, correlating different hadronic decay modes which are sensitive to isospin violation, is only possible if hadronic matrix elements are calculated from first principles like in the framework of QCDF. A method based on flavour symmetries, as it has been used in most studies of $B \to K\pi$ decays so far, could not achieve this. In particular, the decays $B_s \to \phi\pi, \phi\rho$ which are our main interest are not related to any other decay via $SU(3)$ flavour and so their branching fractions could not be predicted in this way.

This chapter is organised as follows: In Section 8.1 we introduce our parametrisation of the Wilson coefficients and discuss the size of the NP parameters which is needed in order to solve the ΔA_{CP} discrepancy. Furthermore, from our QCDF results we derive simple approximate formulae for $\text{Br}(B_s \to \phi\pi)$ and $\text{Br}(B_s \to \phi\rho)$ which permit the calculation of these quantities without implementing the whole apparatus of QCDF. In Section 8.2 we briefly explain the Rfit method [82] before applying it to the $B \to K\pi$ observables and discussing the results. The consequences on the decays $B_s \to \phi\pi, \phi\rho$ are eventually investigated in Section 8.3.

8.1. Modified EW penguin coefficients

In the SM the Wilson coefficients $C_7, ..., C_{10}$ obey at the electroweak scale the hierarchy $C_9 \gg C_7 \gg C_8, C_{10}$. This is because C_9 receives $1/\sin^2\theta_W$-enhanced contributions from Z-penguin and box diagrams in contrast to C_7 while $C_{8,10}$ are generated for the first time at the two-loop level due to their colour structure. For our model independent analysis we include arbitrary NP contributions into the LO coefficients $C_7^{(0)}$ and $C_9^{(0)}$ as well as into their mirror counterparts.

Normalising the new coefficients to the SM value of $C_9^{(0)}$, we have

$$\begin{aligned} C_9^{(0)} &= C_9^{(0),\,\mathrm{SM}}\,(1+\eta_9)\,, & C_9^{(0)\prime} &= C_9^{(0),\,\mathrm{SM}}\,\eta_9'\,,\\ C_7^{(0)} &= C_7^{(0),\,\mathrm{SM}} + C_9^{(0),\,\mathrm{SM}}\,\eta_7\,, & C_7^{(0)\prime} &= C_9^{(0),\,\mathrm{SM}}\,\eta_7'\,. \end{aligned} \tag{8.1}$$

at the weak scale with $\eta_{7,9}^{(\prime)} = q_{7,9}^{(\prime)}\, e^{i\varphi_{7,9}^{(\prime)}}$ providing a new weak phase $\varphi_{7,9}^{(\prime)}$. The coefficient $C_9^{(0),\,\mathrm{SM}}$ contains m_t^2/M_W^2- and $1/\sin^2\theta_W$- enhanced parts of C_9^{SM} as explained in Section 2.1.2. In that section we described also the scheme which we use for the renormalisation group evolution. Applying it to the NP coefficients leads to the low-scale values displayed in table 8.1.

In our analyses we will study several different scenarios. First, we consider the cases where only one of the coefficients η_7, η_9, η_7', η_9' is different from zero. This means we assume the dominance of an individual NP operator as it has also been done for example in Ref. [81]. Second, we consider the possibility of having $\eta_7 = \eta_9$ or $\eta_7' = \eta_9'$. Such a structure would for example arise from NP contributions to photon penguin diagrams. Finally, we study left-right symmetric new contributions corresponding to the three cases $\eta_7 = \eta_7'$, $\eta_9 = \eta_9'$ and $\eta_7 = \eta_7' = \eta_9 = \eta_9'$. Each of these scenarios can be described by means of two real parameters, the absolute value q and phase φ of the NP contributions. This reduced number of free parameters allows us to perform a fit to $B \to K\pi$ data and to draw meaningful conclusion on the $B_s \to \phi\pi, \phi\rho$ decays. The study of this large set of well-motivated simplified scenarios is assumed to represent all relevant features of the general framework with unrelated η_7, η_9, η_7', η_9'.

Our main motivation for adding NP to the coefficients $C_7^{(\prime)}, \ldots, C_{10}^{(\prime)}$ was the claim that the ΔA_{CP} discrepancy can be solved in this way, namely by generating the terms $\tilde{r}_{\mathrm{EW}}$, $\tilde{r}_{\mathrm{EW}}^{\mathrm{C}}$, $\tilde{r}_{\mathrm{EW}}^{\mathrm{A}}$ introduced in (6.13). Setting

$$-\delta \equiv \varphi_7 = \varphi_9 = \varphi_7' = \varphi_p', \tag{8.2}$$

as it is fulfilled for all the scenarios which we consider, we obtain

$$\begin{aligned} \tilde{r}_{\mathrm{EW}} &= (q_7 - q_7')\,\left[(-0.12)^{+0.04}_{-0.05} + (-0.02)^{+0.07}_{-0.02}\, i\right] + \\ &\quad\ (q_9 - q_9')\,\left[0.12^{+0.05}_{-0.04} + 0.02^{+0.02}_{-0.07}\, i\right] \\ \tilde{r}_{\mathrm{EW}}^{\mathrm{C}} &= (q_7 - q_7')\,\left[0.10^{+0.03}_{-0.02} + 0.01^{+0.01}_{-0.06}\, i\right] + \\ &\quad\ (q_9 - q_9')\,\left[0.04^{+0.02}_{-0.03} + (-0.005)^{+0.016}_{-0.026}\, i\right] \\ \tilde{r}_{\mathrm{EW}}^{\mathrm{A}} &= (q_7 - q_7')\,\left[0.03^{+0.04}_{-0.07} + (-0.06)^{+0.12}_{-0.01}\, i\right] + \\ &\quad\ (q_9 - q_9')\,\left[0.007^{+0.003}_{-0.010} + (-0.006)^{+0.012}_{-0.003}\, i\right]\,. \end{aligned} \tag{8.3}$$

Let us briefly discuss the main characteristics of these coefficients:

- First of all, note that left-right symmetric models obviously do not contribute to $B \to K\pi$

	$C_i^{\rm NP}(m_b)/\alpha_e$	$C_i^{{\rm NP}\,\prime}(m_b)/\alpha_e$
C_7	$-0.966\,\eta_7+0.009\,\eta_9$	$-0.966\,\eta_7'+0.009\,\eta_9'$
C_8	$-0.387\,\eta_7+0.002\,\eta_9$	$-0.387\,\eta_7'+0.002\,\eta_9'$
C_9	$0.010\,\eta_7-1.167\,\eta_9$	$0.010\,\eta_7'-1.167\,\eta_9'$
C_{10}	$-0.001\,\eta_7+0.268\,\eta_9$	$-0.001\,\eta_7'+0.268\,\eta_9'$

Table 8.1: NLO Electroweak penguin short-distance coefficients at the scale m_b. Modifications to other short-distance coefficients are negligible.

at all. This general feature of PP decays follows from Eq. (2.14). Therefore such a scenario cannot solve the $\Delta A_{\rm CP}$ discrepancy.

- The $q_7^{(\prime)}$ and $q_9^{(\prime)}$ contributions to $\tilde{r}_{\rm EW}$ tend to cancel each other. Hence in the scenarios with $\eta_7=\eta_9$ or $\eta_7'=\eta_9'$ the coefficient $\tilde{r}_{\rm EW}$ is negligible.
- Whereas the contribution from $q_9^{(\prime)}$ to $\mathrm{Re}(\tilde{r}_{\rm EW}^{\rm C})$ shows the typical colour-suppression compared to the one to $\mathrm{Re}(\tilde{r}_{\rm EW})$, this pattern is not obeyed by the $q_7^{(\prime)}$ terms. This is due to a conspirative interplay of the large mixing of C_7 into C_8 (compare Tab. 8.1), constructive summation of the $C_7/3$ and C_8 contributions to a_8 in Eq. (2.11) and the chiral enhancement factor $r_\chi^{\pi,K}\approx 1.5$ in Eq. (2.12). None of these three effects is present in the q_9 case.
- The annihilation coefficient $\tilde{r}_{\rm EW}^{\rm A}$ develops for $q_7^{(\prime)}\neq 0$ a large imaginary part. In scenarios with non-vanishing $\eta_7^{(\prime)}$ this term gives the dominant contribution to $\Delta A_{\rm CP}$.

From Eq. (6.20) we see that the $\Delta A_{\rm CP}$ discrepancy can be solved either through $\tilde{r}_{\rm EW}$ or through $\tilde{r}_{\rm EW}^{\rm A}$. Except for the left-right symmetric models, all the scenarios mentioned above can achieve such a solution. In Fig. 8.1 this is illustrated for the cases with a single η_7 or η_9 and for the $\eta_7=\eta_9$ scenario. Graphs for the respective mirror scenarios are obtained by a $180°$ rotation. The light-coloured region contains those points of the $(\mathrm{Re}(\eta_i),\mathrm{Im}(\eta_i))$ - plane for which the theory error band overlaps with the experimental $1\,\sigma$ region, whereas the dark-coloured region represents those points for which also the experimental mean value lies within the theory error interval. The circle illustrates the minimal q - value needed to reduce the $\Delta A_{\rm CP}$ tension below the $1\,\sigma$ level. For the three scenarios in Fig. 8.1 we read off $q_7\gtrsim 0.3$, $q_9\gtrsim 0.8$ and $q_7=q_9\gtrsim 0.4$. The fact that in the $\eta_7=\eta_9$ case only a small NP contribution is needed, in spite of the absence of $\tilde{r}_{\rm EW}$, demonstrates the importance of the annihilation term $\tilde{r}_{\rm EW}^{\rm A}$. Finally, we like to stress that the solution of the $\Delta A_{\rm CP}$ discrepancy via a minimal q - value requires the adjustment of the phase φ to a certain value. Realistic scenarios avoiding such a fine-tuning have therefore at least slightly larger q - values, typically $q\sim 1$.

Our main goal is to study the impact of such a NP scenario on the decays $B_s\to\phi\pi,\phi\rho$. The

NP contributions to $C_7^{(\prime)}, ..., C_{10}^{(\prime)}$ generate the $\tilde{r}_{\rm EW}^{\pi/\rho}$ - term introduced in Eq. (7.6). Assuming again universality (8.2) for the weak phases, we obtain

$$\begin{aligned} \tilde{r}_{\rm EW}^{\pi} &= -0.9\ (q_7 + q_7' - q_9 - q_9') \\ \tilde{r}_{\rm EW}^{\rho,0} &= 0.9\ (q_7 - q_7' + q_9 - q_9') \\ \tilde{r}_{\rm EW}^{\rho,-} &= -0.6\ (q_7 + q_9) \\ \tilde{r}_{\rm EW}^{\rho,+} &= 0.6\ (q_7' + q_9') \times P_{\rm EW}^{\rho,-}/P_{\rm EW}^{\rho,+}\,, \end{aligned} \tag{8.4}$$

where we have neglected $q_{7,9}$ contributions to $\tilde{r}_{\rm EW}^{\rho,+}$ and $q_{7,9}'$ contributions to $\tilde{r}_{\rm EW}^{\rho,-}$ according to their $\Lambda_{\rm QCD}^2/m_b^2$ suppression. The SM EW penguin amplitude $P_{\rm EW}^{\rho,+}$ drops out from the total expression (7.6) of the amplitude, $P_{\rm EW}^{\rho,-}$ is given in Eq. (7.5). The parameters $\tilde{r}_{\rm EW}^{\pi/\rho}$ develop only very small imaginary parts and uncertainties not indicated in (8.4). This is because they are ratios of equal topologies such that uncertainties and strong phases approximately cancel. We have stated the expression (8.4) for two reasons: First, in order to make obvious the main consequences of $q_{7,9}^{(\prime)}$ on the $B \to \phi\pi, \phi\rho$ decays. We see that for $q_i = \mathcal{O}(1)$ indeed new contributions with the magnitude of the leading SM EW penguin are generated. While left-right symmetric NP was invisible in $B \to K\pi$, it could be detected in $B_s \to \phi\pi$ and in principle also in $B_s \to \phi\rho$ due to the different interference patterns of $\tilde{r}_{\rm EW}^{\rho,-}$ and $\tilde{r}_{\rm EW}^{\rho,+}$ with the corresponding SM contributions. Furthermore, left- and righthanded NP could be distinguished by a polarisation measurement of $B_s \to \phi\rho$. This general feature of VV decays has been pointed out by Kagan [4]. Note that the question of left- vs. right-handed NP can not be answered from $B \to K\pi$ alone since, as we have seen, the two scenarios differ only by a rotation in the NP parameter space. The second benefit of expression (8.4) is that it allows the calculation of the $B_s \to \phi\pi, \phi\rho$ branching ratios to a very good accuracy. Therefore it permits the study of these decays without the extensive implementation of QCDF. To calculate the branching ratios one simply evaluates Eq. (7.6) inserting (7.4), (7.5) and (8.4). Different phases $\varphi_{7,9}^{(\prime)}$ can be accounted for by individuating the single terms in (8.4). Adding the uncertainties of $P_{\rm EW}^{\pi/\rho}$ and $r_{\rm C}^{\pi/\rho}$ in quadrature leads moreover to a reasonable error estimate.

8.2. Fit to $B \to K\pi$ data

In this section we fit the NP parameters $q_{7,9}^{(\prime)}$, $\varphi_{7,9}^{(\prime)}$ in our various scenarios to the $B \to K\pi$ data. We start by giving a brief review of the *R*fit method [82] which we use.

8.2.1. The *R*fit method

In order to find out which values of the free parameters q_i of a model are most likely to be realised in nature one proceeds as follows: One calculates a set of observables $x_k^{\rm th}(q_1, q_2, ...)$ for

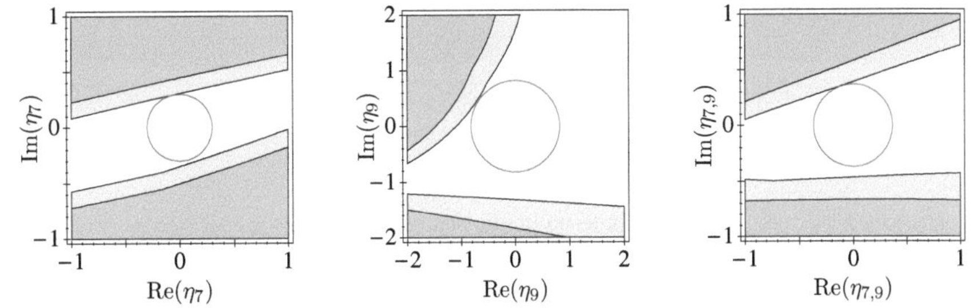

Figure 8.1: NP contribution needed to solve the $\Delta A_{\rm CP}$ discrepancy in the three scenarios (from left to right) with single η_7, single η_9 and equal $\eta_7 = \eta_9$ contribution. Light-coloured region: Theory error band and experimental $1\,\sigma$ region overlap. Dark-coloured region: Theory error band and experimental mean value overlap. Circle: Minimal magnitude of the NP contribution needed to reduce the $\Delta A_{\rm CP}$ discrepancy below the $1\,\sigma$ level.

fixed values of the q_i. Assuming that the x_k obtained in this way are the true values of these observables, the probability for obtaining a certain set of results from the measurements is then determined by statistics. A scenario in which the actually measured values $x_k^{\rm exp}$ occur with a larger probability is then believed to be realised more likely than one which predicts a lower probability. The "best" values for $q_1, q_2, ...$ are therefore those which maximise the probability as a function of the q_i, the so-called likelihood function $\mathcal{L}(q_1, q_2, ...)$, or equivalently minimise the χ^2 function

$$\chi^2_{\rm exp} = \sum_k \frac{(x_k - x_k^{\rm exp})^2}{(\sigma_k^{\rm exp})^2}\,, \tag{8.5}$$

with

$$x_k = x_k^{\rm th}(q_1, q_2, ...). \tag{8.6}$$

Here $\sigma_i^{\rm exp}$ represents the experimental $1\,\sigma$ uncertainty.

The χ^2 formula (8.5) follows from the assumption of a Gaussian distribution for the results of a measurement. The non-trivial task of the analysis is the implementation of the theoretical error into (8.5). If one would know the probability distribution $P\left(x_k | x_k^{\rm th}(q_1, q_2, ...)\right)$ for the theory predictions one could include it into a total χ^2 function which then would have to be minimised by the true set of values $(q_1, q_2, ...; x_1, x_2, ...)$. Note that the true value of the observable x_k is not given anymore by an exact theory value as in (8.6) but must be determined from the fit as well. In practice, however, one hardly knows the distribution of the theory errors since they are not caused by statistics. The best one can usually do is to estimate a range of values $\left[x_k^{\rm th} - \sigma_{k,-}^{\rm th}, x_k^{\rm th} + \sigma_{k,+}^{\rm th}\right]$ within which the true value is expected to lie. The *R*fit scheme in which the true value x_k is assumed to lie certainly within the error interval but no preference is given to any of the allowed values corresponds to a frequentist approach and can formally be implemented by the theoretical

χ^2 function [82]

$$\chi^2_{\text{th}} = \sum_k \begin{cases} 0, & \text{if } \left(x_k^{\text{th}}(q_1, q_2, ...) - \sigma_{k,-}^{\text{th}}\right) \leq x_k \leq \left(x_k^{\text{th}}(q_1, q_2, ...) + \sigma_{k,+}^{\text{th}}\right) \\ \infty, & \text{otherwise} \end{cases} . \tag{8.7}$$

The total $\chi^2 = \chi^2_{\text{exp}} + \chi^2_{\text{th}}$ is then to be minimised by finding appropriate $(q_1, q_2, ...; x_1, x_2, ...)$. After performing the minimisation with respect to the x_k one is left with a χ^2 function for the q_i which reads

$$\chi^2 = \sum_k \begin{cases} \dfrac{\left[\left(x_k^{\text{th}}(q_1, q_2, ...) - \sigma_{k,-}^{\text{th}}\right) - x_k^{\text{exp}}\right]^2}{(\sigma_k^{\text{exp}})^2}, & \text{if } x_k^{\text{exp}} < \left(x_k^{\text{th}}(q_1, q_2, ...) - \sigma_{k,-}^{\text{th}}\right), \\ \dfrac{\left[\left(x_k^{\text{th}}(q_1, q_2, ...) + \sigma_{k,+}^{\text{th}}\right) - x_k^{\text{exp}}\right]^2}{(\sigma_k^{\text{exp}})^2}, & \text{if } x_k^{\text{exp}} > \left(x_k^{\text{th}}(q_1, q_2, ...) + \sigma_{k,+}^{\text{th}}\right), \\ 0, & \text{otherwise} \end{cases} . \tag{8.8}$$

This amounts to choosing x_k from the theory interval $\left[x_k^{\text{th}} - \sigma_{k,-}^{\text{th}}, x_k^{\text{th}} + \sigma_{k,+}^{\text{th}}\right]$ in such a way that it is closest to the experimental x_k^{th} and yields therefore the most conservative χ^2 values.

Using the χ^2 function (8.8) it is possible to define confidence levels by means of the function [82]

$$\begin{aligned} \text{CL}(q_1, q_2, ...) &= \frac{1}{\sqrt{2^{N_{\text{dof}}}}\,\Gamma(N_{\text{dof}}/2)} \int_{\Delta\chi^2(q_1,q_2,...)}^{\infty} e^{-t/2} t^{N_{\text{dof}}/2-1}\, dt\,, \\ \text{with} \quad \Delta\chi^2(q_1, q_2, ...) &= \chi^2(q_1, q_2, ...) - \chi^2_{\text{min}}\,. \end{aligned} \tag{8.9}$$

Here N_{dof} is the number of free model parameters q_i, χ^2_{min} is the minimum of the χ^2 function and Γ denotes the Gamma function. Setting $\text{CL} = 1 - 68.27/100$, $\text{CL} = 1 - 95.45/100$ and $\text{CL} = 1 - 99.73/100$ one individuates the $1\,\sigma$, $2\,\sigma$ and $3\,\sigma$ confidence levels, respectively.

8.2.2. Results

In our χ^2 function we include ΔA_{CP}, ΔA^0_{CP}, the mixing-induced CP asymmetry $S_{K\pi}$ and the one of the pairs (R_c^B, R_n^B), (R_c^K, R_n^K) and (R_c^π, R_n^π) which gives the best constraints for the scenario under consideration. Since hadronic uncertainties partially cancel in $\Delta A^{(0)}_{\text{CP}}$ and in the ratios $R_{c,n}^{B,K,\pi}$, we obtain better results in this way than from fitting directly to branching ratios and CP asymmetries. Apart from the χ^2 fit, we consider constraints at the $2\,\sigma$ level from all the $B \to K\pi$ observables, from the corresponding observables of the decays $\bar{B} \to K\rho$, $K^\star\pi$, $K^\star\rho$ and from other hadronic B decays like $B \to \phi K$, ϕK^*, $\phi\phi$. Especially the constraints from $\bar{B} \to K\rho$ and $\bar{B} \to K^\star\pi$ give some complementary information because they test different chirality structures than $\bar{B} \to K\pi$ and are therefore sensitive to other linear combinations of the

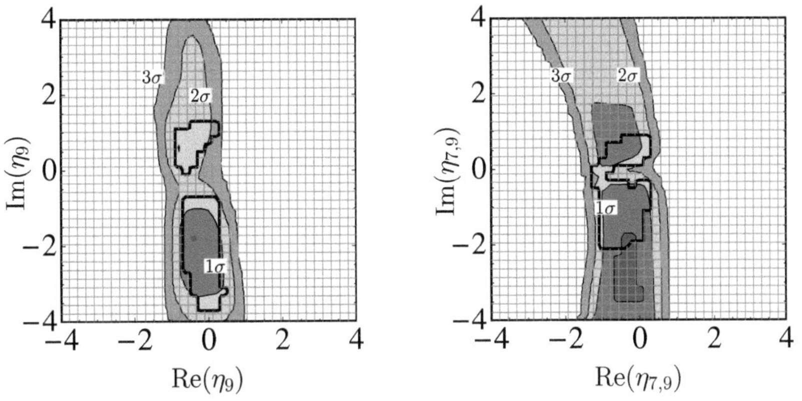

Figure 8.2: Fit to $B \to K\pi$ data for the single η_9 model (left) and the model with $\eta_7 = \eta_9$ (right). For further explanations see text.

$\eta^{(\prime)}_{7,9}$. The resulting fits for the single η_9 and for the $\eta_7 = \eta_9$ scenarios are shown in Fiq. 8.2. The region excluded by the $2\,\sigma$ constraints is hatched by the grey grid. The $1\,\sigma$, $2\,\sigma$ and $3\,\sigma$ regions resulting from the $\bar{B} \to K\pi$ fit are marked by the respective labels and the best fit points are represented by the dark-coloured within 1σ-region. In the single η_9 case the best fit point is given by

$$\hat{q}_9 = 1.84, \qquad \hat{\varphi}_9 = -103°, \tag{8.10}$$

whereas in the $\eta_7 = \eta_9$ case a plateau of $\chi^2 = 0$ points arises due to the large theoretical errors. It turns out that the $B \to K\pi$ observables are not very sensitive to the single η_7 scenario and so the fit does not work well here. Hence within the single η_7 setting one can only rely on the $2\,\sigma$ constraints. The same is of course true for the left-right symmetric models which do not affect $B \to K\pi$ at all.

The fits for the mirror scenarios are simply obtained through a rotation by $180°$. However, this does not hold for the constraints since the $B \to K\rho$, $K^*\pi$ decays are unaffected by a $\eta_{7,9} \to \eta'_{7,9}$ replacement. It turns out that the constraints are stronger in the single η'_9 and in the $\eta'_7 = \eta'_9$ scenarios than in their unprimed counterparts and that the best fit regions are cut away in these cases. The corresponding plots are shown in Fig. 8.3. We further remark that in all the considered scenarios including the single $\eta^{(\prime)}_7$ and the left-right symmetric ones, NP effects are limited to $q_i < 5$ by the $2\,\sigma$ constraints.

In our sample models we introduced NP exclusively in the EW penguin operators. In realistic models, however, a new contribution in the EW penguin sector comes usually in combination with NP of comparable size in the QCD penguins since the new contribution in general matches

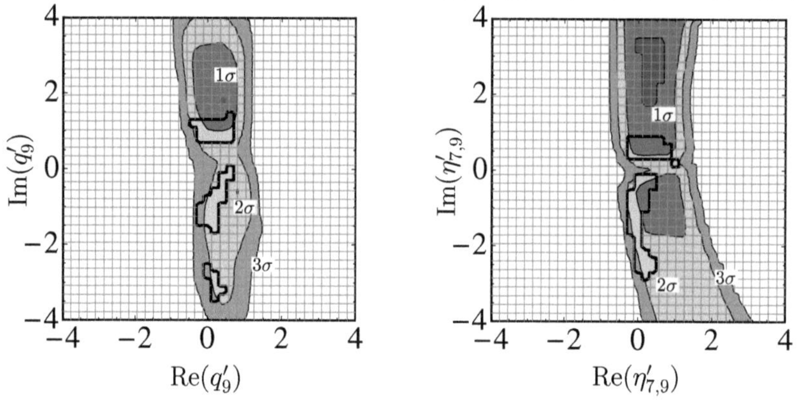

Figure 8.3: Fit to $B \to K\pi$ data for the single η_9' model (left) and the model with $\eta_7' = \eta_9'$ (right). For further explanations see text.

onto a linear combination of the QCD and EW penguin operators. By using mainly isospin violating observables for the fit and as constraints, we reduced the sensitivity to additional effects from $Q_3, ..., Q_6$. Yet in the $\varphi_i^{(\prime)} \approx 90°, 270°$ region, the terms linear in $\tilde{r}_{\rm EW}^{(\rm C)}$ in the $R_{c,n}^{B,K,\pi}$ tend to vanish. The quadratic terms on the other hand contain also interference terms of the new EW penguin contribution with a potential new QCD penguin contribution. Moreover, the quantities $\Delta A_{\rm CP}$ and $\Delta A_{\rm CP}^0$ which are primarily sensitive to isospin violation in the $\varphi_i^{(\prime)} \approx 90°, 270°$ sector are not much constraining (except for excluding $\eta_i^{(\prime)} \approx 0$). Therefore this region is mainly constrained by quadratic contributions to $R_{c,n}^{B,K,\pi}$ and by the mixing induced CP asymmetries $S_{K\pi}$, $S_{\phi K}$ which are sensitive to QCD penguins with a new weak phase as well. Our results for $\varphi_i^{(\prime)} \approx 90°, 270°$ are thus strictly valid only in the pure scenario with NP exclusively in the coefficients $C_7^{(\prime)}, ..., C_{10}^{(\prime)}$ and can be transferred to a more general case only as a order-of-magnitude estimation.

8.3. Consequences for $B \to \phi\pi, \phi\rho$

With the $B \to K\pi$ fits at hand we are now in a position to study a potential enhancement of $\mathrm{Br}(\bar{B}_s \to \phi\pi, \phi\rho)$. To this end we consider all points in parameter space which lie within the $1\,\sigma$ region of the $B \to K\pi$ fit and fulfill the additional $2\,\sigma$ constraints. Our results are displayed in Tab. 8.2. We present the maximum enhancement factor $\mathrm{Br}^{\rm SM+NP}/\mathrm{Br}^{\rm SM}$ of the branching fractions in the various scenarios for $\phi\pi$, $\phi\rho$ and longitudinally polarised $\phi_L\rho_L$ final states. The value for $\mathrm{Br}^{\rm SM}$ is given by the respective mean value in Eqs. (7.9), (7.10) whereas $\mathrm{Br}^{\rm SM+NP}$ is evaluated at the upper end of the error band, i.e. it includes a theory error in favour of an enhancement. Results

Scenario	$\bar{B}_s \to \phi\pi$		$\bar{B}_s \to \phi\rho$		$\bar{B}_s \to \phi_L\rho_L$	
SM	1.7	(1.0)	1.6	(1.0)	1.7	(1.0)
η_7	46.7	(28.6)	23.1	(14.1)	25.0	(14.1)
η_9	15.9	(8.5)	14.7	(8.4)	16.6	(9.0)
$\hat{\eta}_9$	4.5	(2.1)	4.4	(2.4)	4.7	(2.2)
$\eta_7 = \eta_9$	1.7	(1.0)	21.1	(12.6)	23.2	(12.9)
η_7'	59.9	(37.2)	54.4	(34.6)	62.1	(38.6)
η_9'	8.4	(5.3)	3.0	(1.6)	3.2	(1.5)
$\eta_7' = \eta_9'$	1.7	(1.0)	8.0	(5.2)	9.0	(5.7)
$\eta_7 = \eta_7'$	142.1	(87.1)	1.6	(1.0)	1.7	(1.0)
$\eta_9 = \eta_9'$	53.7	(33.4)	1.6	(1.0)	1.7	(1.0)
$\eta_7 = \eta_7' = \eta_9 = \eta_9'$	1.7	(1.0)	1.6	(1.0)	1.7	(1.0)

Table 8.2: Maximum enhancement $\mathrm{Br}^{\mathrm{SM+NP}}/\mathrm{Br}^{\mathrm{SM}}$ in the various scenarios for $\phi\pi$, $\phi\rho$ and longitudinally polarised $\phi_L\rho_L$ final states. The branching ratio $\mathrm{Br}^{\mathrm{SM+NP}}$ is evaluated at the upper end of the theory error band, the result for the mean value is given in brackets.

which are obtained using the mean value instead are given in brackets. Concerning the left-right symmetric scenarios one should have in mind, that like the SM they violate ΔA_{CP} at the $> 2\,\sigma$ level since they have no impact on $B \to K\pi$ decays. The corresponding enhancement factors shown in Tab. 8.2 are obtained by ignoring ΔA_{CP} and considering only all the other constraints. The $\hat{\eta}_9$ scenario in Tab. 8.2 corresponds to the best fit point (8.10) obtained for the single η_9 set-up.

In order to be distinguishable from the SM, a particular scenario must at least provide a value for $\mathrm{Br}^{\mathrm{SM+NP}}/\mathrm{Br}^{\mathrm{SM}}$ which exceeds a potential enhancement factor faked by hadronic uncertainties in the SM prediction and stated in the first line of Tab. 8.2. This is possible in most of the scenarios since typically an enhancement of more than an order of magnitude is allowed. Exceptions are $\bar{B}_s \to \phi\pi$ for $\eta_7^{(\prime)} = \eta_9^{(\prime)}$ and $\bar{B}_s \to \phi_{(L)}\rho_{(L)}$ for left-right symmetric models and have their origin in the pattern of Eq. (8.4). Furthermore, effects in the single η_9' and the $\eta_7' = \eta_9'$ scenarios are limited by the small allowed region resulting from the $B \to K\pi$ fit (compare Fig. 8.3). Largest effects occur as expected in the scenarios which are least constrained by $B \to K\pi$, i.e. the single $\eta_7^{(\prime)}$ and the left-right symmetric models. Especially in these cases a $\bar{B}_s \to \phi\pi$ measurement would complement $B \to K\pi$ data and, while the left-right symmetric models lack the motivation via the ΔA_{CP} discrepancy, the η_7' setting resolves it with ease (see Fig. 8.1).

We have seen that NP in the EW penguin coefficients allows for an enhancement of $\mathrm{Br}(\bar{B}_s \to \phi\pi, \phi\rho)$ of more than an order of magnitude. According to the simple topological structure of

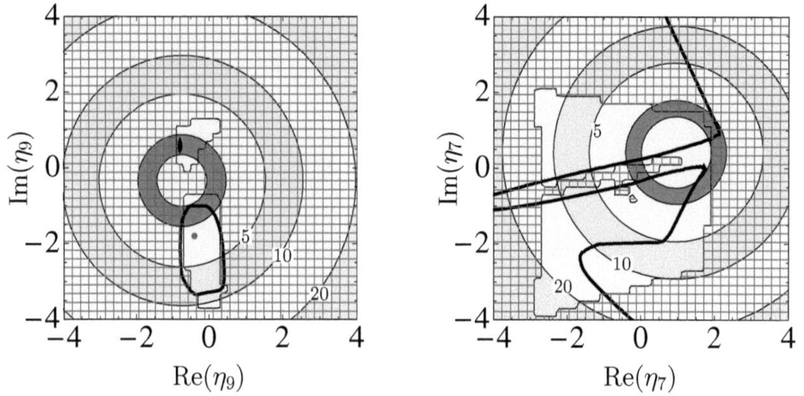

Figure 8.4: Enhancement $\mathrm{Br}^{\mathrm{SM+NP}}/\mathrm{Br}^{\mathrm{SM}}$ for $\bar{B}_s \to \phi\rho$ in the single η_9 scenario (left) and for $\bar{B}_s \to \phi\pi$ in the single η_7 scenario (right). For further explanations see text.

these decays, the observation of such an effect would be a clear and unambiguous signal for such a scenario. It is interesting to raise also the reversed question, i.e. whether the absence of such an effect would rule out a NP solution of the ΔA_{CP} discrepancy, at least for a specific scenario. This is, however, not compulsory. In nearly all the considered settings there are points within the $1\,\sigma$ region of the $B \to K\pi$ fit which do not generate an enhancement of the $\mathrm{Br}(\bar{B}_s \to \phi\pi, \phi\rho)$. The only exception is the single q_7' case: Here an enhancement factor of at least 2.6 would occur in $\bar{B}_s \to \phi\pi$. This time we have exploited the theoretical error in disfavour of an enhancement (for the mean value a factor of 3.3 occurs).

For illustrative purposes we present in Fig. 8.4 a plots of $\mathrm{Br}^{\mathrm{SM+NP}}/\mathrm{Br}^{\mathrm{SM}}$ for $\bar{B}_s \to \phi\rho)$ with the η_9 set-up and for $\bar{B}_s \to \phi\pi$ in the η_7 scenario. The region excluded by the $2\,\sigma$ constraints is once again branded by the grey grid, the $1\,\sigma$ region of the $B \to K\pi$ fit resides inside the thick black curve (in the η_7 case to the left of the curve). The parameter points for which the enhancement factor lies within the theory error band of the SM prediction is displayed by the dark ring.

9. SURVEY OF VIABLE NP MODELS

In the previous chapter we have studied $B \to K\pi$ and $B_s \to \phi\pi, \phi\rho$ in a model-independent way by introducing generic NP contributions to the Wilson coefficients $C_7^{(\prime)}, ..., C_{10}^{(\prime)}$. We have seen that, depending on the specific scenario, NP coefficients with the magnitude of the SM coefficient C_9 can solve the $\Delta A_{\rm CP}$ discrepancy in $B \to K\pi$ and may lead to an order of magnitude enhancement of the branching fractions of $B_s \to \phi\pi, \phi\rho$. Given this situation, it remains to check whether realistic NP models can provide such contributions to the EW penguin coefficients. Furthermore, effects in the non-leptonic decays $B \to K\pi$ and $B_s \to \phi\pi, \phi\rho$ which are generated in a specific model are accompanied by and related to effects in other processes like $B \to X_s\ell^+\ell^-$ and B_s-$\bar{B}_s$-mixing. These processes usually give tight constraints on the new flavour structures and it has to be investigated if the effects in $B \to K\pi$ and $B_s \to \phi\pi, \phi\rho$ survive these constraints.

We start this chapter by listing the processes which set the most stringent constraints. Then we will study in detail the generic situation of a model with a modified flavour-changing Z-coupling as well as a model with an additional $U(1)$ gauge symmetry.

9.1. Constraints from $\bar{B} \to X_s\ell^+\ell^-$, $\bar{B} \to K^*\ell^+\ell^-$ and B_s-$\bar{B}_s$ mixing

The semileptonic decay $\bar{B} \to X_s\ell^+\ell^-$ can be accounted for by enlarging the effective Hamiltonian (2.1) by adding the operators

$$Q_{9V} = \frac{1}{2}(\bar{s}_\alpha\gamma^\mu P_L b_\alpha)(\bar{\ell}\gamma_\mu\ell) \qquad \text{and} \qquad Q_{10A} = \frac{1}{2}(\bar{s}_\alpha\gamma^\mu P_L b_\alpha)(\bar{\ell}\gamma_\mu\gamma_5\ell) \tag{9.1}$$

and their mirror copies Q'_{9V} and Q'_{10A}. The SM expressions for the short-distance coefficients C_{9V} and C_{10A} can be found in Refs. [83,84]. For the renormalisation group evolution we proceed in analogy to the case of the electromagnetic penguin operators, treating those parts of C_{9V} and C_{10A} which are enhanced by $x_{tW} = m_t^2/M_W^2$ and/or $1/\sin^2\theta_W$ as leading order. This results in the following initial conditions at the scale $\mu = \mathcal{O}(M_W)$:

$$\begin{aligned}
C_{9V}^{(0)} &= \frac{\alpha}{2\pi}\left(\frac{Y_0(x_{tW})}{\sin^2\theta_W} - \frac{x_{tW}}{2}\right), \\
C_{9V}^{(1)} &= \frac{\alpha}{2\pi}\left(-4Z_0(x_{tW}) + \frac{x_{tW}}{2} + \frac{4}{9}\right) + \frac{\alpha}{2\pi}\frac{\alpha_s}{4\pi}\left(\frac{Y_1(x_{tW})}{\sin^2\theta_W} - 4x_{tW}\left(\frac{4}{3} - \frac{\pi^2}{6}\right)\right), \\
C_{10A}^{(0)} &= -\frac{\alpha}{2\pi}\frac{Y_0(x_{tW})}{\sin^2\theta_W}, \qquad C_{10V}^{(1)} = -\frac{\alpha}{2\pi}\frac{\alpha_s}{4\pi}\frac{Y_1(x_{tW})}{\sin^2\theta_W}.
\end{aligned} \tag{9.2}$$

The functions $Y_{0,1}$ and Z_0 can be found e.g. in Ref. [85]. Following Ref. [84] and extending the formulae quoted there to include the effects of the mirror operators, we calculate the ratio

$$R_{\ell^+\ell^-}(q^2) \equiv \frac{\frac{d}{dq^2}\Gamma(b \to s\,\ell^+\ell^-)}{\Gamma(b \to c\,\ell\bar{\nu})}, \tag{9.3}$$

where $q^2 = (p_{\ell^+} + p_{\ell^-})^2$ is the invariant massof the lepton pair. We then consider the integrated ratio

$$R_{\ell^+\ell^-}|_{[1,6]} \equiv \int_{1\text{GeV}^2}^{6\text{GeV}^2} R_{\ell^+\ell^-}(q^2)\, dq^2 \tag{9.4}$$

as an additional $2\,\sigma$ constraint with the experimental result given by [86]

$$R_{\ell^+\ell^-}|_{[1,6]} = (1.51 \pm 0.48) \cdot 10^{-5}. \tag{9.5}$$

Apart from the semileptonic inclusive decays $\bar{B} \to X_s\ell^+\ell^-$, the exclusive mode $\bar{B} \to K^*\ell^+\ell^-$ has been proven to be a useful process to constrain new physics, thanks to the possibility of considering various angular observables [87, 88]. Here we focus only on the forward-backward asymmetry $A_{\rm FB}$, which is sufficient to give a constraint complementary to that of $R_{\ell^+\ell^-}|_{[1,6]}$ [87]: We require the sign of $A_{\rm FB}(q^2)$ integrated over $q^2 > 14$ GeV^2 to be negative.

Finally, we consider constraints coming from B_s-$\bar{B}_s$ mixing, which is described by the effective weak Hamiltonian

$$\mathcal{H}_{\rm eff}^{(2)} = \frac{G_F^2 M_W^2}{4\pi^2}(\lambda_t^{(s)})^2 \sum_i C_i Q_i, \tag{9.6}$$

with the operators

$$\begin{aligned}
Q^{\rm VLL} &= (\bar{s}_\alpha\gamma^\mu P_L b_\alpha)(\bar{s}_\beta\gamma_\mu P_L b_\beta), \\
Q_1^{\rm SLL} &= (\bar{s}_\alpha P_L b_\alpha)(\bar{s}_\beta P_L b_\beta), \qquad Q_2^{\rm SLL} = (\bar{s}_\alpha\sigma^{\mu\nu} P_L b_\alpha)(\bar{s}_\beta\sigma_{\mu\nu} P_L b_\beta), \\
Q_1^{\rm LR} &= (\bar{s}_\alpha\gamma^\mu P_L b_\alpha)(\bar{s}_\beta\gamma_\mu P_L b_\beta), \qquad Q_2^{\rm LR} = (\bar{s}_\alpha P_L b_\alpha)(\bar{s}_\beta P_R b_\beta).
\end{aligned} \tag{9.7}$$

In the SM only $C^{\rm VLL} \neq 0$, while in extensions of the SM all operators can receive contributions.

The matrix element relevant for B_s-$\bar{B}_s$ mixing,

$$M_{12}^{B_s} = \frac{1}{2m_{B_s}} \langle B_s^0 | \mathcal{H}_{\text{eff}}^{(2)} | \bar{B}_s^0 \rangle, \tag{9.8}$$

is evaluated using lattice results from Ref. [89]. Besides the B_s-$\bar{B}_s$ mass difference [90]

$$\Delta M_s = 2|M_{12}^{B_s}| \overset{\text{exp.}}{=} (17.77 \pm 0.12)\text{ps}^{-1}, \tag{9.9}$$

we use the quantity

$$\Delta_s \equiv \frac{M_{12}^{B_s}}{M_{12}^{B_s,\text{SM}}} = |\Delta_s| e^{i\phi_s}, \tag{9.10}$$

as additional constraint. This observable has been analysed in Ref. [91] in different generic NP scenarios and the results suggest the possibility of having a NP contribution with a large new weak phase. A fit of Δ_s and the analogous quantity Δ_d to the data shows a $3.4\,\sigma$ discrepancy for the SM value $\Delta_s = 1$. In our study of the Z' models we take those points of the NP parameter space as excluded which give a Δ_s outside the $2\,\sigma$ region drawn in Fig. 9 of Ref. [91].

9.2. Flavour-changing Z-boson coupling

A simple way to get a large new contribution to the EW penguin sector is to consider a flavour-changing $Z\bar{s}b$ coupling. Such a coupling would induce new contributions to $C_7^{(\prime)}, ..., C_{10}^{(\prime)}$ via tree-level Z-exchange. A FCNC Z coupling arises for example at tree level in models with an additional generation of exotic quarks transforming in a non-canonical way under $SU(2)_L$ [92]. Being non-universal in flavour space, the Z-coupling to quarks is then no longer protected by the GIM mechanism and it is rendered flavour non-diagonal by the CKM rotation. In other NP models a FCNC Z coupling is typically generated through a loop of virtual particles. The fact that the $Z\bar{s}b$ coupling is dimensionless allows in principle for a non-decoupling behaviour as it is exhibited by the SM coupling for $m_t \to \infty$. However, the FCNC Z coupling clearly has to involve a $SU(2)_L \times U(1)_Y$ breaking term and gauge invariance implies a v^2/M_{NP}^2 decoupling with M_{NP} representing the scale of the respective NP model [44]. The impact of a model with a modified Z coupling on $B \to K\pi$ has been studied in Refs. [30,43] using flavour symmetries. We will perform here an updated analysis using QCDF and analysing the consequences for $B_s \to \phi\pi, \phi\rho$.

9.2.1. Effective Hamiltonian

We parametrise the $Z\bar{s}b$ coupling in the Lagrangian as

$$\mathcal{L} \supset -\frac{g}{2\cos\theta_W}\,\bar{s}\gamma^\mu\left[\kappa_L^{sb}\,P_L + \kappa_R^{sb}\,P_R\right] b\,Z_\mu\,. \tag{9.11}$$

This parametrisation follows Ref. [43]. Since the flavour violating couplings are expected to be small, the flavour-diagonal couplings are to leading order the same as in the SM. Matching tree-level diagrams with Z exchange onto the $\Delta B = \Delta S = 1$ effective Hamiltonian adds then new contributions δC_i to the SM Wilson coefficients C_i and generates also coefficients C_i' of the mirror operators. The resulting contributions read at the weak scale

$$\begin{aligned}
\delta C_3 &= \frac{1}{6}\frac{\kappa_L^{sb}}{\lambda_t^{(s)}}\,, & C_5' &= \frac{1}{6}\frac{\kappa_R^{sb}}{\lambda_t^{(s)}}\,,\\
\delta C_7 &= \frac{2}{3}\frac{\kappa_L^{sb}}{\lambda_t^{(s)}}\sin^2\theta_W\,, & C_7' &= -\frac{2}{3}\frac{\kappa_R^{sb}}{\lambda_t^{(s)}}\cos^2\theta_W\,,\\
\delta C_9 &= -\frac{2}{3}\frac{\kappa_L^{sb}}{\lambda_t^{(s)}}\cos^2\theta_W\,, & C_9' &= \frac{2}{3}\frac{\kappa_R^{sb}}{\lambda_t^{(s)}}\sin^2\theta_W\,.
\end{aligned} \tag{9.12}$$

A contribution of the same order as the loop-induced Standard Model one arises for

$$|\kappa_{L/R}^{sb}| \sim |\kappa^{\rm SM}| \equiv \frac{\alpha}{\pi\sin^2\theta_W}\,\lambda_t^{(s)}\,C_0(x_{tW}) \sim 0.00035\,, \tag{9.13}$$

where the loop function $C_0(x)$ can be found e.g. in Ref. [85]. The corresponding values at the low scale m_b are obtained by means of the renormalisation group evolution described in Sect. 2.1.2.

Within the same framework, one obtains new contributions to the short-distance coefficients of the semileptonic operators in (9.1), namely

$$\begin{aligned}
\delta C_{9V} &= -\frac{\kappa_L^{sb}}{\lambda_t^{(s)}}\left(2\sin^2\theta_W - \frac{1}{2}\right), & C_{9V}' &= -\frac{\kappa_R^{sb}}{\lambda_t^{(s)}}\left(2\sin^2\theta_W - \frac{1}{2}\right),\\
\delta C_{10A} &= -\frac{\kappa_L^{sb}}{\lambda_t^{(s)}}\left(\frac{1}{2}\right), & C_{10A}' &= -\frac{\kappa_R^{sb}}{\lambda_t^{(s)}}\left(\frac{1}{2}\right).
\end{aligned} \tag{9.14}$$

Diagrams with Z-exchange contribute also to B_s-$\bar{B}_s$ mixing via the Wilson coefficients

$$\begin{aligned}
\delta C_1^{\rm VLL} &= \frac{4\pi^2}{\sqrt{2}\,G_F M_W^2}\left(\frac{\kappa_L^{sb}}{\lambda_t^{(s)}}\right)^2, & C_1^{\rm VRR} &= \frac{4\pi^2}{\sqrt{2}\,G_F M_W^2}\left(\frac{\kappa_R^{sb}}{\lambda_t^{(s)}}\right)^2,\\
C_1^{\rm LR} &= \frac{8\pi^2}{\sqrt{2}\,G_F M_W^2}\,\frac{\kappa_L^{sb}}{\lambda_t^{(s)}}\,\frac{\kappa_R^{sb}}{\lambda_t^{(s)}}\,.
\end{aligned} \tag{9.15}$$

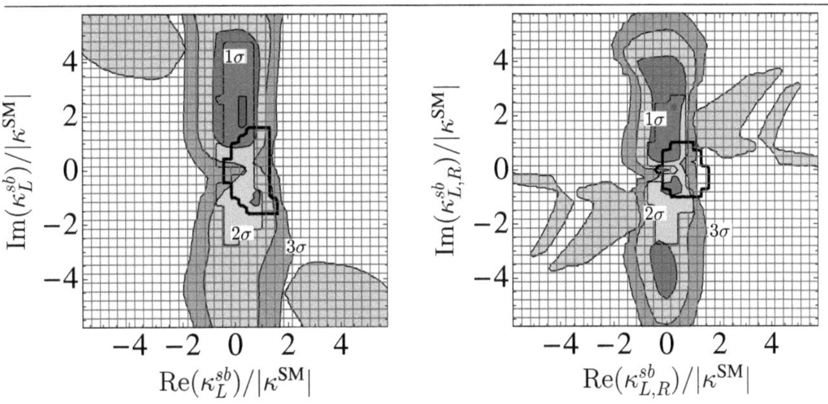

Figure 9.1: Fit to $B \to K\pi$ data for the single κ_L^{sb} model (left) and the model with $\kappa_L^{sb} = \kappa_R^{sb}$ (right). For further explanations see text.

Explaining the Δ_s discrepancy encountered in Sect. 9.1 with the help of these new contributions would push the couplings $\kappa_{L,R}^{sb}$ to large values. Note, however, that in most realistic cases the couplings $\kappa_{L,R}^{sb}$ are loop-induced with the consequence of Eq. (9.15) representing actually two-loop effects. Usually such scenarios provide also one-loop contributions from box diagrams which then are more likely to account for the Δ_s discrepancy. Therefore we prefer not to include Δ_s as a constraint into our analysis and regard a potential relaxation of the Δ_s discrepancy only as a bonus feature.

9.2.2. Results

Similarly to the model-independent analysis, we consider the special cases of a single κ_L^{sb}, a single κ_R^{sb} and $\kappa_L^{sb} = \kappa_R^{sb}$ in our study of the modified Z coupling. Since $\cos^2\theta_W \gg \sin^2\theta_W$, the κ_L^{sb} scenario shares the most important features with the η_9 set-up of the model-independent study and the same holds for κ_R^{sb} and η_7'. The resulting fits for a single κ_L^{sb} and for $\kappa_L^{sb} = \kappa_R^{sb}$, normalised to the SM value κ^{SM}, are shown in Fig. 9.1. The meanings of the coloured regions and the grey grid are the same as in the plots presented in the previous chapter. The main difference to the more general model-independent approach is that we face now additional constraints from semileptonic decays and B_s-$\bar{B}_s$ mixing. The allowed region for the former is given by the interior of the black curve, the allowed region for the latter by the grey areas outside the zone preferred by the $B \to K\pi$ fit. We see that the Δ_s anomaly of B_s-$\bar{B}_s$ mixing cannot be resolved in a modified Z scenario when fulfilling at the same time the semileptonic constraints. This has already been noted in Ref. [93]. Here we recognise that also $B \to K\pi, K\rho, K^*\pi$ data is not compatible with a solution of Δ_s in this way. This statement holds also for the single κ_R^{sb} case not shown in

Scenario	$\bar{B}_s \to \phi\pi$	$\bar{B}_s \to \phi\rho$	$\bar{B}_s \to \phi_L\rho_L$
SM	1.7 (1.0)	1.6 (1.0)	1.7 (1.0)
κ_L^{sb}	2.6 (1.6)	1.8 (1.2)	1.9 (1.2)
κ_R^{sb}	4.0 (2.3)	2.5 (1.5)	2.8 (1.6)
$\kappa_L^{sb} = \kappa_R^{sb}$	1.7 (1.0)	1.6 (1.0)	1.7 (1.0)

Table 9.1: Maximum enhancement $\mathrm{Br}^{\mathrm{SM+NP}}/\mathrm{Br}^{\mathrm{SM}}$ in the various scenarios for $\phi\pi$, $\phi\rho$ and longitudinally polarised $\phi_L\rho_L$ final states. The branching ratio $\mathrm{Br}^{\mathrm{SM+NP}}$ is evaluated at the upper end of the theory error band, the result for the mean value is given in brackets.

Fig. 9.1. Here it is the $2\,\sigma$ constraints from $B \to K\rho, K^*\pi$ which exclude the parameter values required for explaining Δ_s. In the previous section we remarked that it is plausible to assign the explanation of Δ_s to other effects not directly related to the modified Z coupling. Pursuing this strategy, we are left with the semileptonic decays which are compatible with the $1\,\sigma$ region of the $B \to K\pi$ fit for all three cases but constrain the FCNC couplings κ_{LR}^{sb} to very small values.

As a consequence we expect no significant effects in $B_s \to \phi\pi, \phi\rho$. This expectation is confirmed by the maximum enhancement factors given in Tab. 9.1, which are determined in analogy to the ones in Tab. 8.2. In the $\kappa_L^{sb} = \kappa_R^{sb}$ case no enhancement occurs at all because of the pattern in Eq. (8.4): Equal contributions to C_7 and C_9' and to C_9 and C_7' cancel pairwise. The largest effect which one could gain in the other scenarios is a factor of ~ 4 in the case of single κ_R^{sb}. Therefore an enhancement of $B_s \to \phi\pi, \phi\rho$ due to a new modified Z contribution becomes indistinguishable in practise from the potential enhancement caused by a large non-factorisable SM effect. This situation is illustrated in Fig. 9.2 where the solid curve indicates the region allowed from semileptonic decays and the dashed curves surround the $1\,\sigma$ regions of the $B \to K\pi$ fit. We see that the SM error band, represented by the dark ring, nearly fills out the complete allowed region.

Our results can be summarised as follows: The constraints from semileptonic decays still allow for a solution of ΔA_{CP} via a modified Z coupling. This possibility would be excluded if one would find enhancement of $B_s \to \phi\pi$ or $B_s \to \phi\rho$ by an order of magnitude.

9.3. Models with an additional $U(1)$ gauge symmetry

Additional $U(1)$ gauge symmetries are not introduced in order to solve any particular problem of the SM but they appear as remnant of many fundamental NP extensions like Grand Unified Theories, various forms of dynamical symmetry breaking and little Higgs models. The associated Z' gauge bosons are in many scenarios expected to have masses at the TeV scale. An extensive

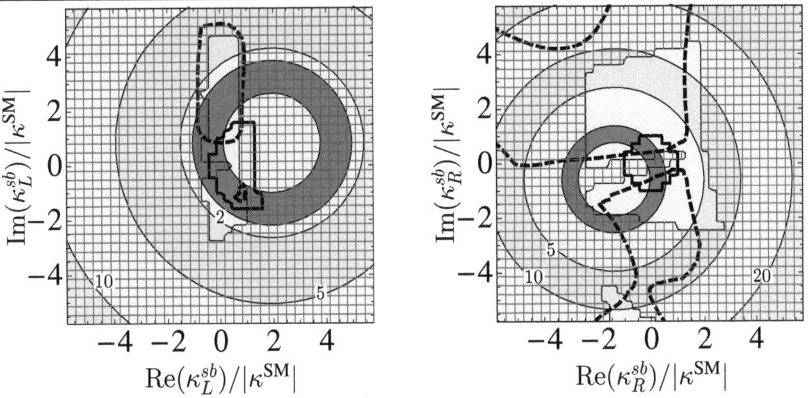

Figure 9.2: Enhancement of $\mathrm{Br}^{\mathrm{SM+NP}}/\mathrm{Br}^{\mathrm{SM}}$ for $\bar{B}_s \to \phi\rho$ in the single κ_L^{sb} scenario (left) and for $\bar{B}_s \to \phi\pi$ in the κ_R^{sb} scenario (right). For further explanations see text.

review about the physics of extra Z' bosons can be found in Ref. [94]. Implications on flavour physics have been discussed in Refs. [45, 95].

9.3.1. Effective Hamiltonian

For simplicity we neglect the effects of Z-Z' mixing and assume the absence of exotic fermions which could mix with the Standard Model fermions through non-universal Z' couplings. We write the general quark-antiquark-Z' boson coupling as [43, 95]

$$\mathcal{L}_{Z'} \supset -\frac{g_{U(1)'}}{\sqrt{2}} \sum_{i,j} \bar{d}_i \gamma^\mu \left[\zeta_L^{ij} P_L + \zeta_R^{ij} P_R\right] d_j \, Z'_\mu \qquad (i,j = d,s,b) \tag{9.16}$$

and similarly for the up-type quarks. The couplings of interest are the flavour-changing $\zeta_{L,R}^{sb}$ as well as the flavour-conserving charges $\zeta_{L,R}^{u} \equiv \zeta_{L,R}^{uu}$ and $\zeta_{L,R}^{d} \equiv \zeta_{L,R}^{dd}$. Note that $SU(2)_L$ invariance implies $\zeta_L^u = \zeta_L^d \equiv \zeta_L^q$ whereas no restrictions hold in case of ζ_R^u, ζ_R^d. Following Ref. [43] we introduce the parameter

$$\xi \equiv \frac{g_{U(1)'}^2}{g^2} \frac{M_W^2}{M_{Z'}^2} \tag{9.17}$$

with $g_{U(1)'}$ denoting the gauge coupling of the additional $U(1)'$ gauge group and $M_{Z'}$ being the mass of the Z'-boson. We find then at the electroweak scale the following additional contributions

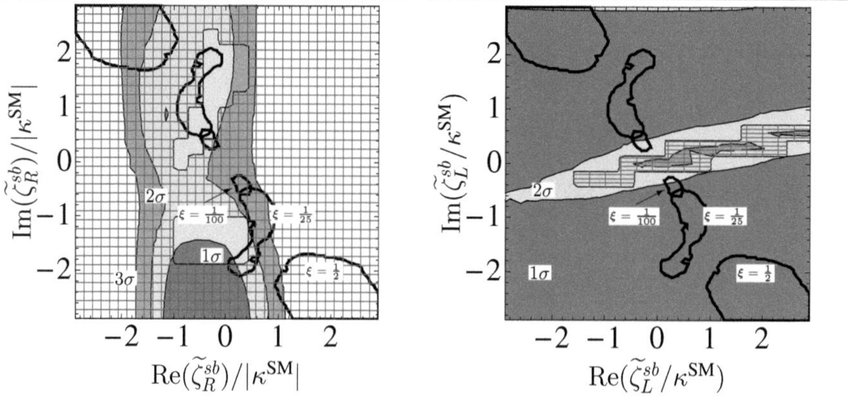

Figure 9.3: Fit to $B \to K\pi$ data for the single ζ_R^{sb} (left) and the single ζ_L^{sb} model (right). Constraints from B_s-$\bar{B}_s$ mixing are displayed for $\xi = 1/2$, $\xi = 1/25$ and $\xi = 1/100$. For further explanations see text.

to the short-distance coefficients in Eq. (2.1):

$$
\begin{aligned}
\delta C_3 &= -\frac{\zeta_L^{sb}}{\lambda_t^{(s)}}\,\zeta_L^q\,\xi\,, & C_3' &= -\frac{1}{3}\frac{\zeta_R^{sb}}{\lambda_t^{(s)}}\left(\zeta_R^u + 2\zeta_R^d\right)\xi\,,\\
\delta C_5 &= -\frac{1}{3}\frac{\zeta_L^{sb}}{\lambda_t^{(s)}}\left(\zeta_R^u + 2\zeta_L^d\right)\xi\,, & C_5' &= -\frac{\zeta_R^{sb}}{\lambda_t^{(s)}}\,\zeta_L^q\,\xi\,,\\
\delta C_7 &= -\frac{2}{3}\frac{\zeta_L^{sb}}{\lambda_t^{(s)}}\left(\zeta_R^u - \zeta_R^d\right)\xi\,, & C_7' &= 0\,,\\
\delta C_9 &= 0\,, & C_9' &= -\frac{2}{3}\frac{\zeta_R^{sb}}{\lambda_t^{(s)}}\left(\zeta_R^u - \zeta_R^d\right)\xi\,.
\end{aligned}
\tag{9.18}
$$

Following the approach of Ref. [95] we assume the largest contributions to reside in the EW penguin coefficients. This can be arranged by choosing $\zeta_L^q = 0$ and $(\zeta_R^u + 2\zeta_R^d) = 0$. We are left then with only one non-zero charge combination $(\zeta_R^u - \zeta_R^d)$ which can be set to $(\zeta_R^u - \zeta_R^d) = 1$ without loss of generality by a proper redefinition of the Abelian coupling $g_{U(1)'}$. The coupling of the Z' boson to quarks is not related to its coupling to leptons. Therefore tight constraints from semileptonic decays, as we encountered in the case of a modified Z coupling, can be avoided here by simply switching off the Z' coupling to leptons. Such "leptophobic" Z' bosons can for example appear in models with a E_6 gauge symmetry (see e.g. Ref. [96]). Since leptophobic Z' bosons avoid detection via traditional Drell-Yan processes, their mass is much less constrained allowing for larger values of the parameter ξ.

Apart from the non-leptonic processes we are interested in, the only constraints come from B_s-$\bar{B}_s$

mixing. The corresponding Wilson coefficients read

$$\delta C_1^{\rm VLL} = \frac{4\pi^2\sqrt{2}}{G_F M_W^2}\left(\frac{\zeta_L^{sb}}{\lambda_t^{(s)}}\right)^2 \xi, \qquad C_1^{\rm VRR} = \frac{4\pi^2\sqrt{2}}{G_F M_W^2}\left(\frac{\zeta_R^{sb}}{\lambda_t^{(s)}}\right)^2 \xi,$$
$$C_1^{\rm LR} = \frac{8\pi^2\sqrt{2}}{G_F M_W^2}\left(\frac{\zeta_L^{sb}}{\lambda_t^{(s)}}\right)\left(\frac{\zeta_R^{sb}}{\lambda_t^{(s)}}\right)\xi. \tag{9.19}$$

9.3.2. Results

We consider the three special cases of single ζ_L^{sb}, single ζ_R^{sb} and $\zeta_L^{sb} = \zeta_R^{sb}$. From Eq. (9.18) we immediately see that the single ζ_L^{sb} scenario is in direct correspondence to the single η_7 case discussed in our model-independent study in Chapter 8. The same holds for ζ_R^{sb} and η_9'. Therefore the results of the $B \to K\pi$ fits presented in Figs. 8.2 and 8.3 can directly be translated into fits for the quantities $\widetilde{\zeta}_{L,R}^{sb} = \xi\,\zeta_{L,R}^{sb}$. To this end the corresponding graphs must be rotated by 180° according to the minus signs in Eq. (9.18) and the normalisation of the axes has to be adjusted properly. As a consequence the enhancement factors of $B_s \to \phi\pi, \phi\rho$ stated in Tab. 8.2 for η_7 and η_9' apply also to the single ζ_L^{sb} and single ζ_R^{sb} scenarios, respectively, as long as only constraints from non-leptonic B decays are considered. For $\zeta_L^{sb} = \zeta_R^{sb}$, on the other hand, neither $B_s \to \phi\pi$ nor $B_s \to \phi\rho$ develop an enhancement because of the cancellation of the η_7 and η_9' effects in (8.4).

As novel feature compared to the model-independent analysis we have additional constraints from B_s-$\bar{B}_s$ mixing. In the relevant Wilson coefficients (9.19) the parameters $\zeta_{L,R}^{sb}$ and ξ enter always in combinations

$$\xi\,\zeta_i^{sb}\,\zeta_j^{sb} = \frac{1}{\xi}\,\widetilde{\zeta}_i^{sb}\,\widetilde{\zeta}_j^{sb}, \qquad i,j = L,R. \tag{9.20}$$

Therefore the B_s-$\bar{B}_s$ mixing constraint in the $(\mathrm{Re}(\widetilde{\zeta}_i^{sb}), \mathrm{Im}(\widetilde{\zeta}_i^{sb}))$ - plane depends on the parameter ξ determined by the coupling constant $g_{U(1)'}$ and the Z' mass $M_{Z'}$. It gets stronger for smaller ξ, i.e. for smaller $g_{U(1)'}$ and larger Z' mass $M_{Z'}$. This behaviour, which might seem counter-intuitive at first sight, has its origin in the dependence of the hadronic decays on the parameter combinations $\widetilde{\zeta}_i^{sb} = \xi\,\zeta_i^{sb}$. If one chooses smaller ξ values, one needs larger values of the FCNC couplings ζ_i^{sb} in order to obtain the same effects in the hadronic decays. Since the B_s-$\bar{B}_s$ mixing coefficients (9.19) depend quadratically on the ζ_i^{sb}, this procedure sharpens their constraints.

In Fig. 9.3 we present our results of the $B \to K\pi$ fits for the single ζ_L^{sb} and the single ζ_R^{sb} scenarios. The 2σ region for Δ_s is shown for different values of ξ. We recognise that there is very little overlap of the $1\,\sigma$ region of the $B \to K\pi$ fit with the region preferred by Δ_s in the ζ_R^{sb} case. The same holds for the $\zeta_L^{sb} = \zeta_R^{sb}$ scenario, not shown in Fig. 9.3. This phenomenon is easily understood: The observables $\Delta A_{\rm CP}$ and Δ_s both call for NP with a large imaginary

Scenario	$\bar{B}_s \to \phi\pi$	$\bar{B}_s \to \phi\rho$	$\bar{B}_s \to \phi_L\rho_L$
SM	1.7 (1.0)	1.6 (1.0)	1.7 (1.0)
ζ_L^{sb}	5.5 (3.5)	4.9 (3.2)	5.5 (3.5)
ζ_R^{sb}	6.5 (4.1)	5.9 (3.8)	6.8 (4.3)
$\zeta_L^{sb} = \zeta_R^{sb}$	1.7 (1.0)	1.6 (1.0)	1.7 (1.0)

Table 9.2: Maximum enhancement $\mathrm{Br}^{\mathrm{SM+NP}}/\mathrm{Br}^{\mathrm{SM}}$ in the various scenarios for $\xi = 1/25$. The branching ratio $\mathrm{Br}^{\mathrm{SM+NP}}$ is evaluated at the upper end of the theory error band, the result for the mean value is given in brackets.

part. The branching ratios of hadronic B decays depend at leading order linearly on the real part of $\zeta_{L,R}^{sb}$ and draw the $\zeta_{L,R}^{sb}$ values therefore to the imaginary axis. The observable Δ_s, on the other hand, depends quadratically on $\zeta_{L,R}^{sb}$ and favours values on the diagonal. For the ζ_L^{sb} setting this situation is relaxed due to the weak constraints from $B \to K\pi$. From the diagrams we see further that the B_s-$\bar{B}_s$ mixing constraint is very tight. It prohibits large effects in $B_s \to \phi\pi, \phi\rho$ for realistic values of the parameter $\xi \lesssim 1/25$. For $\xi = 1/25$, which would correspond for example to $g_{U(1)'} \sim g$ and $M_{Z'} \sim 400\,\mathrm{GeV}$, we present the maximum enhancement factors in Tab. 9.2. These numbers are obtained abandoning the $1\,\sigma$ region of the $B \to K\pi$ fit and requiring only agreement with the $2\,\sigma$ constraints. We find that enhancement of a factor ~ 5 is possible in the ζ_L^{sb} and ζ_R^{sb} scenarios whereas no effect can occur in the $\zeta_L^{sb} = \zeta_R^{sb}$ case because of Eq. (8.4). For $\xi < 1/100$ the constraints from B_s-$\bar{B}_s$ mixing become so strong that practically no effect in $B_s \to \phi\pi, \phi\rho$ would be detectable. Measurement of a significant enhancement would therefore set a lower limit on ξ, equivalent to an upper limit on the Z' mass $M_{Z'}$.

10. Conclusions

In this thesis we have studied the impact of different NP scenarios on rare B decays mediated by a FCNC $b \to s$ transition. Our focus has been on non-leptonic decays, governed by the effective Hamiltonian (2.1), and throughout the work we relied on QCD factorisation (QCDF) [3] for the evaluation of hadronic matrix elements.

The first part of the thesis addressed a particular NP scenario, the MSSM for large values of $\tan\beta$. We considered a version of Minimal Flavour Violation (MFV) in which all elementary couplings of neutral bosons to (s)quarks are flavour-diagonal and the flavour structures of W, charged-Higgs and chargino couplings are determined by the CKM matrix. It is well-known that in this scenario perturbative calculations of Feynman amplitudes have to be supported by a resummation of $\tan\beta$-enhanced higher order corrections. This subject is usually treated with the help of an effective field theory which is found by integrating out the genuine supersymmetric particles and is therefore valid only for $M_{\text{SUSY}} \gg v, M_{A^0,H^0,H^\pm}$. Using the diagrammatic method developed in Ref. [14] and extending it to the case of flavour-changing interactions, we derived resummation formulae which do not assume any hierarchy between M_{SUSY}, the electroweak scale v and the Higgs masses.

As a first result we found that the resummation formula for the Yukawa coupling y_{d_i} of down-type quarks depends on the renormalisation scheme chosen for the MSSM parameters. In particular, the familiar expression of Eq. (4.24) is modified if the sbottom mixing angle $\tilde{\theta}_b$ is used as input. This result is useful in high-p_T collider physics, since it permits the correct treatment of $\tan\beta$-enhanced effects in production and decay of bottom squarks.

Our main achievement is, however, the resummation of $\tan\beta$-enhanced loop corrections to flavour-changing processes for arbitrary values of SUSY masses. We translated the results into effective Feynman rules (collected in Section 4.5.2) which permit the inclusion of the resummed corrections into calculations of Feynman amplitudes beyond the decoupling limit. Complex phases of flavour-conserving parameters like the trilinear SUSY-breaking term A_t are consistently included in our results. The effective Feynman rules can easily be implemented into computer programs like FeynArts [28] and they can be applied to FCNC processes with virtual SUSY particles as well as to collider processes with external SUSY particles.

We found that our results for the renormalisation of CKM elements and for the loop-induced neutral-Higgs couplings to quarks are represented by the same functions in terms of the parameters ϵ_b and $\epsilon_{\rm FC}$ as in the decoupling limit $M_{\rm SUSY} \gg v, M_{A^0,H^0,H^\pm}$. However, the parameters ϵ_b and $\epsilon_{\rm FC}$ as an incarnation of self-energy diagrams now contain also decoupling parts. As novel results we found $\tan\beta$-enhanced loop-induced couplings of gluinos and neutralinos and determined the analogous corrections to chargino couplings. These results permit the calculation of $\tan\beta$-enhanced corrections to processes involving a decoupling supersymmetric loop which cannot be studied consistently with the effective-field-theory method.

We then analysed the impact of the new FCNC gluino couplings on rare B decays. To this end we determined analytic expressions for all gluino contributions to the effective Hamiltonian (2.1), including gluon-, photon- and Z-penguin as well as box diagrams (collected in Appendix A.4). Most of these contributions turned out to be numerically small for two reasons: Firstly, for positive μ and $\tan\beta \sim 50$ typical values of the FCNC gluino coupling are $\sim\ 0.1 \ll 1$. Secondly, the gluino contributions suffer from GIM-like suppression effects. There is one exception: Chirally enhanced contributions to the magnetic and chromomagnetic operators $Q_{7\gamma}$ and Q_{8g} involve a left-right flip in the squark line proportional to the corresponding quark mass und thus avoid the GIM cancellation. Whereas the contribution from gluino-squark loops to $C_{7\gamma}$ is accidentally small, the one to C_{8g} can indeed contribute as much as the chargino-squark diagram. In order to illustrate the phenomenological consequences of this gluino-squark contribution to C_{8g}, we briefly discussed its effect on the mixing-induced CP asymmetry in the decay $B^0 \to \phi K_S$. In this context we improved the leading order analyses performed in Refs. [74,75] by our full NLO QCDF treatment.

In the second part of the thesis we have studied the possibility of probing isospin violation in hadronic B decays. We started by examining the situation in $B \to K\pi$ decays in light of the current data. Using QCDF for our analysis, the only observable which we found in disagreement with its SM prediction is $\Delta A_{\rm CP}$. Here, our QCDF result $\Delta A_{\rm CP} \stackrel{\rm SM}{=} 1.9^{+5.8}_{-4.8}\%$ deviates from the measured value $\Delta A_{\rm CP} \stackrel{\rm exp.}{=} (14.8 \pm 2.8)\%$ by $\sim 2.5\,\sigma$. We demonstrated in a model-independent analysis that this discrepancy can easily be resolved by an additional NP contribution to the EW penguin operators $Q_7^{(\prime)}, ..., Q_{10}^{(\prime)}$ which is of the same order as the SM coefficient $C_9^{\rm SM}$. An exception are left-right symmetric scenarios where the contributions to PP decays cancel. Furthermore we pointed out that, in the case of NP in $C_7^{(\prime)}$ the solution comes about via an annihilation contribution in the QCDF framework, whereas the solution in the case of NP in $C_9^{(\prime)}$ is, as expected, due to a new contribution to the colour-allowed EW penguin amplitude. For various scenarios we performed frequentist fits to $B \to K\pi$ data. We found the fit to work good for NP in $C_9^{(\prime)}$ while NP in $C_7^{(\prime)}$ is only poorly constrained from $B \to K\pi$ alone. Especially in this case, the PV counterparts $B \to K\rho$ and $B \to K^*\pi$, which are sensitive to a different chirality structure than $B \to K\pi$, give valuable additional information.

Motivated by the $\Delta A_{\rm CP}$ discrepancy, we suggested the decays $B_s \to \phi\pi, \phi\rho$ as golden channels for the study of isospin violation. These decays are purely isospin-violating and dominated by the EW penguin topology. In this work we performed the first analysis of the impact of NP in EW penguins on these decays. From our full QCDF results we derived simple approximate expressions (Eqs. (7.4)-(7.6), (8.4)) which reproduce the $B_s \to \phi\pi, \phi\rho$ amplitudes with high accuracy for arbitrary scenarios with NP in $C_7^{(\prime)}, ..., C_{10}^{(\prime)}$. By quoting these formulae we facilitate the study of these decays without an extensive implementation of the QCDF framework. With respect to our $B \to K\pi$ fits we investigated then for various scenarios the maximum enhancement of the $B_s \to \phi\pi, \phi\rho$ branching ratios. The results displayed in Tab. 8.2 show that in many cases an enhancement by an order of magnitude is possible. Particular exceptions are left-right symmetric models which have no impact on the VV decay $B_s \to \phi\rho$ and scenarios with (approximately) equal contributions to $C_7^{(\prime)}$ and $C_9^{(\prime)}$ which cancel in $B_s \to \phi\pi$.

Finally, we analysed some models in which the new contributions in the EW penguin sector are generated via exchange of the SM Z-boson or of an additional Z'-boson with flavour-changing couplings. A flavour-changing Z-couling is induced by penguin diagrams in the SM and it can for example receive additional contributions from loops involving new particles, like in supersymmetry. We found that in such a scenario the semileptonic constraints still allow for NP to an extent which is sufficient to resolve the $\Delta A_{\rm CP}$ discrepancy. On the other hand, they prevent the $B_s \to \phi\pi, \phi\rho$ decays from developing an enhancement which beats the hadronic uncertainties of the SM prediction. Therefore, a large effect measured in these decays would rule out the modified Z coupling. The semileptonic constraints can for example be avoided in a model with an additional Z' boson whose couplings to leptons can be switched off. Such a "leptophobic" Z' can appear for example in models with a E_6 gauge symmetry. Our analysis showed that in this scenario constraints from hadronic B decays and B_s-$\bar{B}_s$ mixing can be fulfilled simultaneously only at the $2\,\sigma$ level. The tight constraints from B_s-$\bar{B}_s$ mixing limit a potential enhancement of $B_s \to \phi\pi, \phi\rho$ to a factor ~ 5 in the case of a large $g_{U(1)'}$ coupling and a light Z' boson.

We stress again that the decays $B_s \to \phi\pi, \phi\rho$ are highly sensitive to isospin-violating NP. Their measurement would complement the analysis of $B \to K\pi$ decays and could shed light on the "$\Delta A_{\rm CP}$ puzzle". For this reason we like to encourage experimental efforts to measure these decays at LHCb and at prospective Super-B factories.

A. APPENDIX

A.1. Sparticle mixing

In this section our conventions for sparticle masses and mixing matrices are defined. We follow closely the conventions of the SLHA [61] which we extend in Section A.1.1 to account for complex phases in the squark mass matrices. In Section A.1.3 we derive explicit expressions for certain combinations of elements of the chargino mixing matrices.

A.1.1. Squark mixing

In the naive MFV scenario the squark mass-matrices are block-diagonal in the Super-CKM basis with one hermitian 2×2-block for each squark flavour $\tilde{q}$. Let us parametrise these blocks as

$$M_{\tilde{q}}^2 = \begin{pmatrix} m_{\tilde{q}_L}^2 & X_{\tilde{q}} \\ X_{\tilde{q}}^* & m_{\tilde{q}_R}^2 \end{pmatrix}. \tag{A.1}$$

The real diagonal elements are given by

$$\begin{aligned} m_{\tilde{q}_L}^2 &= \widetilde{m}_Q^2 + m_q^2 + (T_q^3 - Q_q \sin^2\theta_W) M_Z^2 \cos 2\beta, \\ m_{\tilde{q}_R}^2 &= \widetilde{m}_q^2 + m_q^2 + Q_q \sin^2\theta_W M_Z^2 \cos 2\beta. \end{aligned} \tag{A.2}$$

with $\widetilde{m}_Q^2$ and $\widetilde{m}_q^2$ being the respective entries of the diagonal soft matrices $(\widetilde{m}_Q^2)^{ij}$ and $(\widetilde{m}_{u,d}^2)^{ij}$ introduced in Eq. (3.6). The expressions m_q, T_q^3 and Q_q denote the mass, weak isospin and charge of the corresponding partner-quark, M_Z and θ_W are the Z-boson mass and the weak mixing angle. The off-diagonal elements read

$$X_{\tilde{u}} = y_u (A_u^* v_u - \mu v_d), \qquad X_{\tilde{d}} = y_d^{(0)*} (A_d^* v_d - \mu v_u), \tag{A.3}$$

for up- and down-type quarks, respectively. Here the entries of the diagonal matrices $a_{u,d}^{ij}$ defined in Eq. (3.6) have been parametrised as $a_q = A_q y_q$. The Yukawa coupling $y_d^{(0)}$ contains $\tan\beta$-enhanced counterterm contributions as discussed in Section 4.3[1].

[1]The corresponding corrections to m_d^2 in the diagonal elements of the squark mass-matrix are negligible since $m_d^2 \ll \widetilde{m}_{Q_L}^2, \widetilde{m}_{d_R}^2$.

Performing unitary rotations

$$\widetilde{R}^q = \begin{pmatrix} \cos\tilde{\theta}_q & \sin\tilde{\theta}_q e^{i\tilde{\phi}_q} \\ -\sin\tilde{\theta}_q e^{-i\tilde{\phi}_q} & \cos\tilde{\theta}_q \end{pmatrix} \tag{A.4}$$

on the quark fields we can diagonalise the mass matrices as follows:

$$\widetilde{R}^q M_{\tilde{q}}^2 \widetilde{R}^{q\dagger} = \mathrm{diag}(m_{\tilde{q}_1}^2, m_{\tilde{q}_2}^2), \tag{A.5}$$

$$m_{\tilde{q}_{1,2}}^2 = \frac{1}{2}\left(m_{\tilde{q}_L}^2 + m_{\tilde{q}_R}^2 \pm \sqrt{(m_{\tilde{q}_L}^2 - m_{\tilde{q}_R}^2)^2 + 4|X_q|^2}\right). \tag{A.6}$$

The mixing-angle $\tilde{\theta}_q$ and the phase $\tilde{\phi}_q$ can be expressed by means of the relation

$$e^{i\tilde{\phi}_q}\sin 2\tilde{\theta}_q = \frac{2X_{\tilde{q}}}{m_{\tilde{q}_1}^2 - m_{\tilde{q}_2}^2}. \tag{A.7}$$

Note that for the determination of the masses from Eq. (A.6) and the mixing parameters from Eq. (A.7) an iterative proceeding might be necessary depending on the renormalisation scheme (see discussion in Section 4.3.3). To give separate expressions for $\tilde{\theta}_q$ and $\tilde{\phi}_q$ one has to specify the allowed range for both parameters. Choosing $\tilde{\theta}_q \in [0, \pi/4]$ and $\tilde{\phi}_q \in [0, 2\pi)$ for example results in

$$\sin 2\tilde{\theta}_q = \left|\frac{2X_{\tilde{q}}}{m_{\tilde{q}_1}^2 - m_{\tilde{q}_2}^2}\right|, \qquad \tilde{\phi}_q = \arg\left(\frac{2X_{\tilde{q}}}{m_{\tilde{q}_1}^2 - m_{\tilde{q}_2}^2}\right). \tag{A.8}$$

Constraining $\tilde{\theta}_q$ to this interval amounts to defining $\tilde{q}_1$ ($\tilde{q}_2$) as the eigenstate which is predominantly left-handed (right-handed).

A.1.2. Chargino mixing

In our conventions the chargino mass-matrix is given by

$$M_{\tilde{\chi}^\pm} = \begin{pmatrix} M_2 & \sqrt{2}M_W\sin\beta \\ \sqrt{2}M_W\cos\beta & \mu \end{pmatrix}. \tag{A.9}$$

We define the biunitary transformation which brings it into diagonal form as

$$\widetilde{U}^* M_{\tilde{\chi}^\pm} \widetilde{V}^\dagger = \mathrm{diag}(m_{\tilde{\chi}_1^\pm}, m_{\tilde{\chi}_2^\pm}). \tag{A.10}$$

The matrices $\widetilde{U}$ and $\widetilde{V}$ are chosen in such a way that $m_{\tilde{\chi}_{1,2}^\pm}$ are real an positive. They can be determined by diagonalising the matrices $M_{\tilde{\chi}^\pm}^\dagger M_{\tilde{\chi}^\pm}$ and $M_{\tilde{\chi}^\pm} M_{\tilde{\chi}^\pm}^\dagger$, respectively. In Feynman amplitudes for diagrams with chirality-flipping propagators only certain combinations of matrix-elements of $\widetilde{U}$ and $\widetilde{V}$ appear. These combinations can be expressed as

$$\begin{aligned}
\widetilde{U}_{11}\widetilde{V}_{11} &= \frac{m_{\widetilde{\chi}_1^\pm}M_2 - m_{\widetilde{\chi}_2^\pm}\mu^*\, e^{i\psi}}{m^2_{\widetilde{\chi}_1^\pm} - m^2_{\widetilde{\chi}_2^\pm}}, & \widetilde{U}_{11}\widetilde{V}_{12} &= \sqrt{2}M_W \frac{m_{\widetilde{\chi}_1^\pm}\sin\beta + m_{\widetilde{\chi}_2^\pm}\cos\beta\, e^{i\psi}}{m^2_{\widetilde{\chi}_1^\pm} - m^2_{\widetilde{\chi}_2^\pm}}, \\
\widetilde{U}_{12}\widetilde{V}_{12} &= \frac{m_{\widetilde{\chi}_1^\pm}\mu - m_{\widetilde{\chi}_2^\pm}M_2^*\, e^{i\psi}}{m^2_{\widetilde{\chi}_1^\pm} - m^2_{\widetilde{\chi}_2^\pm}}, & \widetilde{U}_{12}\widetilde{V}_{11} &= \sqrt{2}M_W \frac{m_{\widetilde{\chi}_1^\pm}\cos\beta + m_{\widetilde{\chi}_2^\pm}\sin\beta\, e^{i\psi}}{m^2_{\widetilde{\chi}_1^\pm} - m^2_{\widetilde{\chi}_2^\pm}}, \\
\widetilde{U}_{21}\widetilde{V}_{21} &= \frac{m_{\widetilde{\chi}_1^\pm}\mu^*\, e^{i\psi} - m_{\widetilde{\chi}_2^\pm}M_2}{m^2_{\widetilde{\chi}_1^\pm} - m^2_{\widetilde{\chi}_2^\pm}}, & \widetilde{U}_{21}\widetilde{V}_{22} &= -\sqrt{2}M_W \frac{m_{\widetilde{\chi}_1^\pm}\cos\beta\, e^{i\psi} + m_{\widetilde{\chi}_2^\pm}\sin\beta}{m^2_{\widetilde{\chi}_1^\pm} - m^2_{\widetilde{\chi}_2^\pm}}, \\
\widetilde{U}_{22}\widetilde{V}_{22} &= \frac{m_{\widetilde{\chi}_1^\pm}M_2^*\, e^{i\psi} - m_{\widetilde{\chi}_2^\pm}\mu}{m^2_{\widetilde{\chi}_1^\pm} - m^2_{\widetilde{\chi}_2^\pm}}, & \widetilde{U}_{22}\widetilde{V}_{21} &= -\sqrt{2}M_W \frac{m_{\widetilde{\chi}_1^\pm}\sin\beta\, e^{i\psi} + m_{\widetilde{\chi}_2^\pm}\cos\beta}{m^2_{\widetilde{\chi}_1^\pm} - m^2_{\widetilde{\chi}_2^\pm}}
\end{aligned} \tag{A.11}$$

with

$$e^{i\psi} = (M_2\mu - M_W^2 \sin 2\beta)/(m_{\widetilde{\chi}_1^\pm} m_{\widetilde{\chi}_2^\pm}). \tag{A.12}$$

For large $\tan\beta$ the $\cos\beta$-terms can be neglected and the above expressions reduce to

$$\begin{aligned}
\widetilde{U}_{11}\widetilde{V}_{11} &= \frac{M_2}{m_{\widetilde{\chi}_1^\pm}} \cdot \frac{m^2_{\widetilde{\chi}_1^\pm} - |\mu|^2}{m^2_{\widetilde{\chi}_1^\pm} - m^2_{\widetilde{\chi}_2^\pm}}, & \widetilde{U}_{11}\widetilde{V}_{12} &= \frac{\sqrt{2}M_W m_{\widetilde{\chi}_1^\pm}\sin\beta}{m^2_{\widetilde{\chi}_1^\pm} - m^2_{\widetilde{\chi}_2^\pm}}, \\
\widetilde{U}_{12}\widetilde{V}_{12} &= \frac{\mu}{m_{\widetilde{\chi}_1^\pm}} \cdot \frac{m^2_{\widetilde{\chi}_1} - |M_2|^2}{m^2_{\widetilde{\chi}_1^\pm} - m^2_{\widetilde{\chi}_2^\pm}}, & \widetilde{U}_{12}\widetilde{V}_{11} &= \frac{M_2}{m_{\widetilde{\chi}_1^\pm}} \cdot \frac{\sqrt{2}M_W \mu\sin\beta}{m^2_{\widetilde{\chi}_1^\pm} - m^2_{\widetilde{\chi}_2^\pm}}, \\
\widetilde{U}_{21}\widetilde{V}_{21} &= \frac{M_2}{m_{\widetilde{\chi}_2^\pm}} \cdot \frac{|\mu|^2 - m^2_{\widetilde{\chi}_2^\pm}}{m^2_{\widetilde{\chi}_1^\pm} - m^2_{\widetilde{\chi}_2^\pm}}, & \widetilde{U}_{21}\widetilde{V}_{22} &= -\frac{\sqrt{2}M_W m_{\widetilde{\chi}_2^\pm}\sin\beta}{m^2_{\widetilde{\chi}_1^\pm} - m^2_{\widetilde{\chi}_2^\pm}}, \\
\widetilde{U}_{22}\widetilde{V}_{22} &= \frac{\mu}{m_{\widetilde{\chi}_2^\pm}} \cdot \frac{|M_2|^2 - m^2_{\widetilde{\chi}_2^\pm}}{m^2_{\widetilde{\chi}_1^\pm} - m^2_{\widetilde{\chi}_2^\pm}}, & \widetilde{U}_{22}\widetilde{V}_{21} &= -\frac{\mu}{m_{\widetilde{\chi}_2^\pm}} \cdot \frac{\sqrt{2}M_W M_2\sin\beta}{m^2_{\widetilde{\chi}_1^\pm} - m^2_{\widetilde{\chi}_2^\pm}}.
\end{aligned} \tag{A.13}$$

A.1.3. Neutralino mixing

In our conventions the neutralino mass-matrix is given by

$$M_{\widetilde{\chi}^0} = \begin{pmatrix} M_1 & 0 & -M_Z\sin\theta_W\cos\beta & M_Z\sin\theta_W\sin\beta \\ 0 & M_2 & M_Z\cos\theta_W\cos\beta & -M_Z\cos\theta_W\sin\beta \\ -M_Z\sin\theta_W\cos\beta & M_Z\cos\theta_W\cos\beta & 0 & -\mu \\ M_Z\sin\theta_W\sin\beta & -M_Z\cos\theta_W\sin\beta & -\mu & 0 \end{pmatrix}. \tag{A.14}$$

We define the unitary transformation which brings it into diagonal form as

$$\widetilde{N}^* M_{\widetilde{\chi}^0} \widetilde{N}^\dagger = \text{diag}(m_{\widetilde{\chi}_1^0}, m_{\widetilde{\chi}_2^0}, m_{\widetilde{\chi}_3^0}, m_{\widetilde{\chi}_4^0}). \tag{A.15}$$

The matrix $\widetilde{N}$ is chosen in such a way that $m_{\widetilde{\chi}^0_{1..4}}$ are real an positive.

A.2. Loop functions

The results for the quark self-energies with internal SUSY particles are given in terms of the following functions:

$$\begin{aligned}
B_0(m_1, m_2) &= \frac{2}{4-d} - \gamma_E + \log 4\pi + 1 - \frac{1}{m_1^2 - m_2^2}\left[m_1^2 \log\frac{m_1^2}{\mu^2} - m_2^2 \log\frac{m_2^2}{\mu^2}\right], \\
C_0(m_1, m_2, m_3) &= \frac{m_2^2}{(m_1^2 - m_2^2)(m_3^2 - m_2^2)} \log\frac{m_1^2}{m_2^2} + \frac{m_3^2}{(m_1^2 - m_3^2)(m_2^2 - m_3^2)} \log\frac{m_1^2}{m_3^2}, \\
D_0(m_1, m_2, m_3, m_4) &= \frac{m_2^2}{(m_2^2 - m_1^2)(m_2^2 - m_3^2)(m_2^2 - m_4^2)} \log\frac{m_1^2}{m_2^2} + \\
&\quad (2 \leftrightarrow 3) + (2 \leftrightarrow 4) \\
D_2(m_1, m_2, m_3, m_4) &= \frac{m_2^4}{(m_2^2 - m_1^2)(m_2^2 - m_3^2)(m_2^2 - m_4^2)} \log\frac{m_1^2}{m_2^2} + \\
&\quad (2 \leftrightarrow 3) + (2 \leftrightarrow 4).
\end{aligned} \tag{A.16}$$

In our expressions for the Wilson coefficients, we use the following loop functions:

- Gluino-induced penguin contributions:

$$\begin{aligned}
f_{8g}^F(x) &= -\frac{x+1}{2(x-1)^2} + \frac{x\log x}{(x-1)^3}, & f_{8g}^A(x) &= -\frac{1}{2(x-1)} + \frac{x\log x}{2(x-1)^2}, \\
f_{7\gamma}^F(x), &= -f_{8g}^F(x)/3 & f_{7\gamma}^A(x) &= 0 \\
\tilde{f}_{8g}^F(x) &= \frac{x^2-5x-2}{12(x-1)^3} + \frac{x\log x}{2(x-1)^4}, & \tilde{f}_{8g}^A(x) &= f_{8g}^F(x)/4, \\
\tilde{f}_{7\gamma}^F(x), &= -\tilde{f}_{8g}^F(x)/3 & \tilde{f}_{7\gamma}^A(x) &= 0 \\
f_{Pg}^F(x) &= \frac{2x^2-7x+11}{18(x-1)^3} - \frac{\log x}{3(x-1)^4}, & f_{Pg}^A(x) &= -\frac{1}{2(x-1)} + \frac{(2x+1)\log x}{6(x-1)^2}, \\
f_{P\gamma}^F(x), &= -f_{Pg}^F(x)/3 & f_{P\gamma}^A(x) &= 0 \\
f_Z(x,y) &= \frac{1}{x-y}\left[\frac{2x^2\log x}{x-1} - \frac{2y^2\log y}{y-1}\right].
\end{aligned} \tag{A.17}$$

- Chargino-induced penguin contributions:

$$g_{7\gamma}(x) = \frac{5-7x}{12(x-1)^2} + \frac{x(3x-2)\log x}{6(x-1)^3}, \qquad g_{8g}(x) = \frac{x+1}{4(x-1)^2} - \frac{x\log x}{2(x-1)^3}. \quad \text{(A.18)}$$

- Box contributions:

$$\begin{aligned} F(x,y) &= -\frac{x\log x}{(x-y)(x-1)^2} - \frac{y\log y}{(y-x)(y_1)^2} - \frac{1}{(x-1)(y-1)}, \\ G(x,y) &= \frac{x^2\log x}{(x-y)(x-1)^2)} + \frac{y^2\log y}{(y-x)(y-1)^2} + \frac{1}{(x-1)(y-1)}. \end{aligned} \quad \text{(A.19)}$$

A.3. Feynman rules for large $\tan\beta$

In this section we give explicit Feynman rules for the flavour-changing vertices generated by replacement rule (iii) in Section 4.5.2. We suppress colour indices of (s)quarks, repeated indices are not summed over.

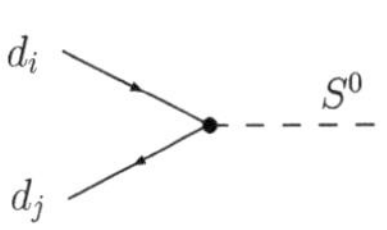

$$-\frac{i}{\sqrt{2}}\left[x_d^S\left(\delta_{ji}\,y_{d_j}^{(0)} + \frac{\delta Z_{ji}^L}{2}\,y_{d_j}^{(0)} - \frac{\delta Z_{ji}^R}{2}\,y_{d_i}^{(0)}\right)P_L \right.$$
$$\left. +(x_d^S)^*\left(\delta_{ji}\,y_{d_j}^{(0)*} + \frac{\delta Z_{ji}^R}{2}\,y_{d_j}^{(0)*} - \frac{\delta Z_{ji}^L}{2}\,y_{d_i}^{(0)*}\right)P_R\right] \quad \text{(A.20)}$$

with $\quad x_d^S = (\cos\alpha, -\sin\alpha, i\sin\beta, -i\cos\beta) \quad$ for $\quad S^0 = (H^0, h^0, A^0, G^0)$

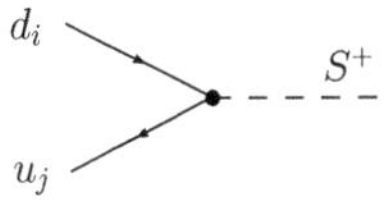

$$i\xi_L^S\,y_{u_j}\,V_{ji}\,P_L + i\xi_R^S\left(y_{d_i}^{(0)*}\,V_{ji}^{(0)} + \frac{\delta Z_{ji}^R}{2}\,y_{d_j}^{(0)*}\,V_{jj}\right)P_R \quad \text{(A.21)}$$

with $\quad \xi_L^S = (\cos\beta, \sin\beta) \quad$ and $\quad \xi_R^S = (\sin\beta, -\cos\beta) \qquad$ for $\quad S^+ = (H^+, G^+)$ (A.22)

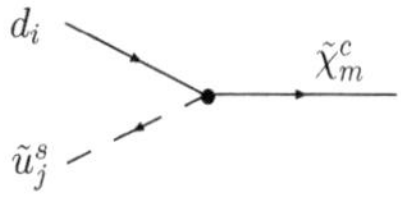

$$iV_{ji}\left(y_{u_j}\widetilde{R}^{u_j}_{s2}\widetilde{V}^*_{m2} - g\widetilde{R}^{u_j}_{s1}\widetilde{V}^*_{m1}\right)P_L + i\widetilde{R}^{u_j}_{s1}\widetilde{U}_{m2}\left(y^{(0)*}_{d_i}V^{(0)}_{ji} + \frac{\delta Z^R_{ji}}{2}y^{(0)*}_{d_j}V_{jj}\right)P_R \tag{A.23}$$

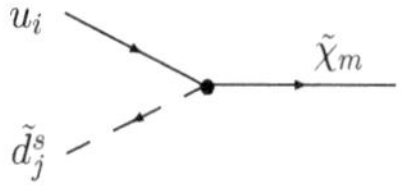

$$iV^{(0)*}_{ij}\left[\left(y^{(0)}_{d_j}\widetilde{R}^{d_j}_{s2}\widetilde{U}^*_{m2} - g\widetilde{R}^{d_j}_{s1}\widetilde{U}^*_{m1}\right)P_L + y_{u_i}\widetilde{R}^{d_j}_{s1}\widetilde{V}_{m2}P_R\right] \tag{A.24}$$

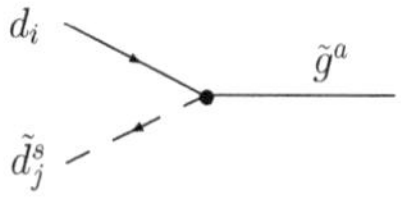

$$-i\sqrt{2}g_sT^a\left[\left(\delta_{ji} + \frac{\delta Z^L_{ji}}{2}\right)\widetilde{R}^{d_j}_{s1}P_L - \left(\delta_{ji} + \frac{\delta Z^R_{ji}}{2}\right)\widetilde{R}^{d_j}_{s2}P_R\right] \tag{A.25}$$

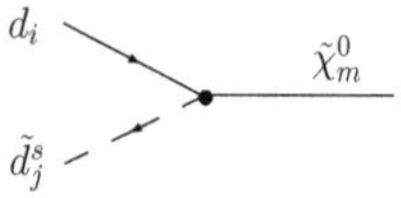

$$i\left(\delta_{ji} + \frac{\delta Z^L_{ji}}{2}\right)\left[\sqrt{2}\widetilde{R}^{d_j}_{s1}\left(\frac{g}{2}\widetilde{N}^*_{m2} - \frac{g'}{6}\widetilde{N}^*_{m1}\right) - y^{(0)}_{d_j}\widetilde{R}^{d_j}_{s2}\widetilde{N}^*_{m3}\right]P_L - i\left(\delta_{ji} + \frac{\delta Z^R_{ji}}{2}\right)\left[\frac{\sqrt{2}}{3}g'\widetilde{R}^{d_j}_{s2}\widetilde{N}_{m1} + y^{(0)*}_{d_j}\widetilde{R}^{d_j}_{s1}\widetilde{N}_{m3}\right]P_R \tag{A.26}$$

A.4. Gluino contributions to the $\Delta B = 1 = \Delta S = 1$ Hamiltonian

In this section we quote our results for the gluino contributions to the Wilson coefficients of the $\Delta B = \Delta S = 1$ Hamiltonian defined in Eqs. (2.1) and (2.2). We will use a notation which allows us to apply the results to the case of naive MFV (as in chapter 5) as well as to the generic MSSM. To this end we numerate the squarks as $\tilde{d}_\alpha$ with $\alpha = 1, ..., 6$ instead of $\tilde{d}_i^s$ with $i = 1, 2, 3$ and $s = 1, 2$ as in the case of naive MFV. The 6×6 mixing matrix $\widetilde{\mathcal{R}}^d$ appearing in the generic MSSM is defined in complete analogy to its 2×2 counterparts $\widetilde{R}^d$ in Eq. (A.5).

The generic form of the quark-squark-gluino coupling in the down-sector is given by

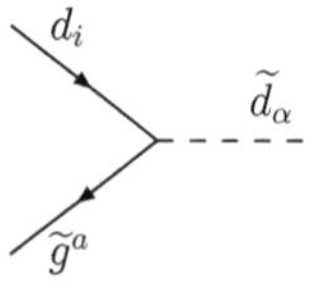

$$-i\sqrt{2}g_sT^A\left[G^L_{\alpha i}P_L - G^R_{\alpha i}P_R\right] \tag{A.27}$$

where the 6×3 matrices $G^L_{\alpha i}$ and $G^R_{\alpha i}$ parametrise the flavour structure. In the generic MSSM they are simply given by

$$G^L_{\alpha i} = \widetilde{\mathcal{R}}^d_{\alpha i}, \qquad G^R_{\alpha i} = \widetilde{\mathcal{R}}^d_{\alpha, i+3}\,. \tag{A.28}$$

In the MSSM with naive MFV they read

$$G^L_{\alpha i} = \begin{cases} \left(\delta_{\alpha i} + \frac{\delta Z^L_{\alpha i}}{2}\right) \widetilde{R}^{d_\alpha}_{11}, & \alpha = 1,2,3 \\ \left(\delta_{\alpha-3,i} + \frac{\delta Z^L_{\alpha-3,i}}{2}\right) \widetilde{R}^{d_{\alpha-3}}_{21}, & \alpha = 4,5,6 \end{cases},$$

$$G^R_{\alpha i} = \begin{cases} \left(\delta_{\alpha i} + \frac{\delta Z^R_{\alpha i}}{2}\right) \widetilde{R}^{d_\alpha}_{12}, & \alpha = 1,2,3 \\ \left(\delta_{\alpha-3,i} + \frac{\delta Z^R_{\alpha-3,i}}{2}\right) \widetilde{R}^{d_{\alpha-3}}_{22}, & \alpha = 4,5,6 \end{cases}. \tag{A.29}$$

We decompose the Wilson coefficients into contributions C_{Pg} from gluon penguins, $C_{P\gamma}$ from photon penguins, κ^{sb}_L from Z penguins and $b^{u,d}_{1,\dots,4}$ from box diagrams:

$$\begin{aligned}
C_3 &= -\frac{1}{6}\frac{\alpha_s}{4\pi} C_{Pg} + \frac{1}{6}\frac{\kappa^{sb}_L}{\lambda^{(s)}_t} + \frac{1}{3}(b^u_3 + 2b^d_3), \\
C_4 &= \frac{1}{2}\frac{\alpha_s}{4\pi} C_{7Pg} + \frac{1}{3}(b^u_4 + 2b^d_4), \\
C_5 &= -\frac{1}{6}\frac{\alpha_s}{4\pi} C_{Pg} + \frac{1}{3}(b^u_1 + 2b^d_1), \\
C_6 &= \frac{1}{2}\frac{\alpha_s}{4\pi} C_{Pg} + \frac{1}{3}(b^u_2 + 2b^d_2), \\
C_7 &= \frac{2}{3}\frac{\alpha_e}{4\pi} C_{P\gamma} + \frac{2}{3}\sin^2\theta_W \frac{\kappa^{sb}_L}{\lambda^{(s)}_t} + \frac{2}{3}(b^u_1 - b^d_1), \\
C_8 &= \frac{2}{3}(b^u_2 - b^d_2), \\
C_9 &= \frac{2}{3}\frac{\alpha_e}{4\pi} C_{P\gamma} - \frac{2}{3}\cos^2\theta_W \frac{\kappa^{sb}_L}{\lambda^{(s)}_t} + \frac{2}{3}(b^u_3 - b^d_3), \\
C_{10} &= \frac{2}{3}(b^u_4 - b^d_4).
\end{aligned} \tag{A.30}$$

The parameter κ^{sb}_L representing the Z penguin is defined in the same way as the modified Z coupling introduced in Chapter 9. Parametrising the quark-squark-gluino vertex as in (A.27) we obtain the following results from squark-gluino loops:

$$\begin{aligned}
C^{\tilde g}_{7\gamma,8g} &= \frac{\sqrt{2}}{4G_F\lambda^{(s)}_t}\frac{g_s^2}{m_{\tilde g}^2}\sum_{\alpha=1}^{6}\left[\frac{m_{\tilde g}}{m_b} G^{L*}_{\alpha 2} G^R_{\alpha 3}\left(C_F f^F_{7\gamma,8g}(x_\alpha) + C_A f^A_{7\gamma,8g}(x_\alpha)\right)\right. \\
&\qquad \left. + G^{L*}_{\alpha 2} G^L_{\alpha 3}\left(C_F \tilde f^F_{7\gamma,8g}(x_\alpha) + C_A \tilde f^A_{7\gamma,8g}(x_\alpha)\right)\right], \\
C^{\tilde g}_{P\gamma,Pg} &= \frac{\sqrt{2}}{4G_F\lambda^{(s)}_t}\frac{g_s^2}{m_{\tilde g}^2}\sum_{\alpha=1}^{6}\left[G^{L*}_{\alpha 2} G^L_{\alpha 3}\left(C_F f^F_{P\gamma,Pg}(x_\alpha) + C_A f^A_{P\gamma,Pg}(x_\alpha)\right)\right],
\end{aligned} \tag{A.31}$$

$$\kappa_{sb}^{L,\tilde{g}} = -\frac{1}{2}\frac{\alpha_s}{4\pi}C_F \sum_{\alpha,\beta=1}^{6}\sum_{i=1}^{3} G_{\alpha 2}^{L*}\, G_{\alpha i}^{R}\, G_{\beta i}^{R*}\, G_{\beta 3}^{L}\, f_Z(x_\alpha, x_\beta), \tag{A.32}$$

Neglecting left-right mixing of the first two squark generations as well as successive $3 \to 1$ and $1 \to 2$ flavour transitions, we obtain for the gluino boxes ($q = u, d$):

$$\begin{aligned}
b_{1q}^{\tilde{g}} &= \frac{\sqrt{2}}{4G_F\lambda_t^{(s)}}\frac{\alpha_s^2}{m_{\tilde{g}}^2}\sum_{\alpha=1}^{6} G_{\alpha 2}^{L*}G_{\alpha 3}^{L}\left[\frac{1}{18}F(x_\alpha, x_{q+3}) - \frac{5}{18}G(x_\alpha, x_{q+3})\right] \\
b_{2q}^{\tilde{g}} &= \frac{\sqrt{2}}{4G_F\lambda_t^{(s)}}\frac{\alpha_s^2}{m_{\tilde{g}}^2}\sum_{\alpha=1}^{6} G_{\alpha 2}^{L*}G_{\alpha 3}^{L}\left[\frac{7}{6}F(x_\alpha, x_{q+3}) + \frac{1}{6}G(x_\alpha, x_{q+3})\right] \\
b_{3q}^{\tilde{g}} &= \frac{\sqrt{2}}{4G_F\lambda_t^{(s)}}\frac{\alpha_s^2}{m_{\tilde{g}}^2}\sum_{\alpha=1}^{6} G_{\alpha 2}^{L*}G_{\alpha 3}^{L}\left[-\frac{5}{9}F(x_\alpha, x_q) + \frac{1}{36}G(x_\alpha, x_q)\right] \\
b_{4q}^{\tilde{g}} &= \frac{\sqrt{2}}{4G_F\lambda_t^{(s)}}\frac{\alpha_s^2}{m_{\tilde{g}}^2}\sum_{\alpha=1}^{6} G_{\alpha 2}^{L*}G_{\alpha 3}^{L}\left[\frac{1}{3}F(x_\alpha, x_q) + \frac{7}{12}G(x_\alpha, x_q)\right].
\end{aligned} \tag{A.33}$$

A.5. QCDF results for $B \to K\rho, K^*\pi, K^*\rho$

In the Tab. A.1 we have collected our SM predictions for the decays $B \to K\rho, K^*\pi, K^*\rho$. The branching ratios and CP asymmetries are abbreviated as Br^{ij}, A_{CP}^{ij} where the first index refers to the charge of the $K^{(*)}$ and the second index to the charge of the π (ρ). The observables $R_{c,n}^{M}$ and $\Delta A_{\mathrm{CP}}^{-,0}$ are defined in analogy to the corresponding quantities for $B \to K\pi$ in chapter 6.

	$B \to K\rho$		$B \to K^*\pi$		$B \to K^*\rho$	
Observable	Theory	Experiment	Theory	Experiment	Theory	Experiment
$\mathrm{Br}^{00} \times 10^6$	$3.1^{+6.1}_{-1.9}$	$4.7^{+0.7}_{-0.7}$	$2.5^{+2.6}_{-2.8}$	$2.4^{+0.7}_{-0.7}$	$2.6^{+2.7}_{-1.8}$	$3.4^{+1.0}_{-1.0}$
$\mathrm{Br}^{-+} \times 10^6$	$4.4^{+9.8}_{-3.1}$	$8.6^{+0.9}_{-1.1}$	$7.7^{+6.0}_{-7.4}$	$8.6^{+0.9}_{-1.0}$	$5.8^{+4.5}_{-2.6}$	< 12
$\mathrm{Br}^{-0} \times 10^6$	$1.8^{+3.7}_{-1.3}$	$3.81^{+0.48}_{-0.46}$	$6.1^{+3.8}_{-5.2}$	$6.9^{+2.3}_{-2.3}$	$4.5^{+2.7}_{-1.8}$	< 6.1
$\mathrm{Br}^{0-} \times 10^6$	$3.8^{+9.6}_{-3.1}$	$8.0^{+1.5}_{-1.4}$	$8.4^{+6.6}_{-8.6}$	$9.9^{+0.8}_{-0.9}$	$6.0^{+5.0}_{-2.9}$	$9.2^{+1.5}_{-1.5}$
R_c^B	$0.95^{+0.53}_{-0.27}$	$0.95^{+0.23}_{-0.19}$	$1.45^{+2.33}_{-0.27}$	$1.39^{+0.48}_{-0.48}$	$1.46^{+0.51}_{-0.25}$	—
R_n^B	$0.71^{+0.21}_{-0.17}$	$0.91^{+0.19}_{-0.17}$	$1.52^{+6.73}_{-0.33}$	$1.79^{+0.76}_{-0.46}$	$1.10^{+0.93}_{-0.35}$	$1.26^{+0.56}_{-0.35}$
$R_c^{K^{(*)}}$	$0.75^{+0.24}_{-0.22}$	$0.82^{+0.16}_{-0.13}$	$1.47^{+0.72}_{-0.17}$	$1.49^{+0.54}_{-0.52}$	$1.46^{+0.52}_{-0.35}$	—
$R_n^{K^{(*)}}$	$0.57^{+0.20}_{-0.23}$	$0.79^{+0.20}_{-0.17}$	$1.54^{+3.22}_{-0.27}$	$1.92^{+0.81}_{-0.47}$	$1.10^{+0.98}_{-0.43}$	$1.26^{+0.56}_{-0.35}$
$R_c^{\pi,\rho}$	$0.80^{+0.20}_{-0.27}$	$0.87^{+0.21}_{-0.17}$	$1.01^{+0.17}_{-0.46}$	$1.07^{+0.17}_{-0.14}$	$1.00^{+0.16}_{-0.18}$	—
$R_n^{\pi,\rho}$	$0.54^{+0.28}_{-0.20}$	$0.75^{+0.16}_{-0.13}$	$2.23^{+15.61}_{-0.67}$	$2.68^{+1.42}_{-1.08}$	$1.61^{+2.24}_{-0.78}$	$1.26^{+0.56}_{-0.35}$
R	$1.20^{+0.34}_{-0.20}$	$1.03^{+0.16}_{-0.14}$	$1.06^{+0.40}_{-0.04}$	$0.99^{+0.26}_{-0.26}$	$1.18^{+0.16}_{-0.08}$	—
A_{CP}^{00}	$-0.22^{+0.32}_{-0.21}$	$0.01^{+0.20}_{-0.20}$	$-0.00^{+0.50}_{-0.64}$	$-0.15^{+0.12}_{-0.12}$	$-0.22^{+0.32}_{-0.21}$	$0.09^{+0.19}_{-0.19}$
A_{CP}^{-+}	$0.39^{+0.20}_{-0.91}$	$0.15^{+0.06}_{-0.06}$	$-0.14^{+0.85}_{-0.50}$	$-0.18^{+0.08}_{-0.08}$	$0.36^{+0.10}_{-0.41}$	—
A_{CP}^{-0}	$0.71^{+0.25}_{-1.65}$	$0.37^{+0.11}_{-0.11}$	$-0.07^{+0.66}_{-0.29}$	$0.04^{+0.29}_{-0.29}$	$0.41^{+0.14}_{-0.40}$	$0.20^{+0.32}_{-0.29}$
A_{CP}^{0-}	$0.001^{+0.013}_{-0.022}$	$-0.12^{+0.17}_{-0.17}$	$0.007^{+0.041}_{-0.006}$	$-0.038^{+0.042}_{-0.042}$	$-0.002^{+0.004}_{-0.017}$	$-0.01^{+0.16}_{-0.16}$
$\Delta A_{\mathrm{CP}}^{-}$	$0.31^{+0.11}_{-0.74}$	$0.22^{+0.13}_{-0.13}$	$0.06^{+0.24}_{-0.22}$	$0.22^{+0.30}_{-0.30}$	$0.05^{+0.16}_{-0.17}$	—
$\Delta A_{\mathrm{CP}}^{0}$	$0.07^{+0.12}_{-0.37}$	$-0.13^{+0.26}_{-0.26}$	$0.007^{+0.649}_{-0.500}$	$0.11^{+0.13}_{-0.13}$	$0.25^{+0.31}_{-0.44}$	$0.10^{+0.25}_{-0.25}$
S_{CP}	$0.46^{+0.21}_{-0.17}$	$0.54^{+0.18}_{-0.21}$	$0.82^{+0.16}_{-0.22}$	—	—	—

Table A.1: Theoretical versus experimental results for the $B \to K\rho, K^*\pi, K^*\rho$ decays. The experimental data is taken from [42].

BIBLIOGRAPHY

[1] L. Hofer, U. Nierste and D. Scherer, JHEP **0910** (2009) 081 arXiv:0907.5408 [hep-ph]; L. Hofer, U. Nierste and D. Scherer, PoS E **PS-HEP2009** (2009) 181 [arXiv:0909.4749 [hep-ph]]; L. Hofer, Acta Phys. Polon. B **3** (2010) 235 [arXiv:0910.0474 [hep-ph]].

[2] L. Hofer, D. Scherer and L. Vernazza, Acta Phys. Polon. B **3** (2010) 227 [arXiv:0910.2809 [hep-ph]]; L. Hofer, D. Scherer, L. Vernazza, JHEP **1102** (2011) 080. [arXiv:1011.6319 [hep-ph]].

[3] M. Beneke, G. Buchalla, M. Neubert and C. T. Sachrajda, Phys. Rev. Lett. **83** (1999) 1914 [arXiv:hep-ph/9905312]; M. Beneke, G. Buchalla, M. Neubert and C. T. Sachrajda, Nucl. Phys. B **591** (2000) 313 [arXiv:hep-ph/0006124]; M. Beneke, G. Buchalla, M. Neubert and C. T. Sachrajda, Nucl. Phys. B **606** (2001) 245 [arXiv:hep-ph/0104110]; M. Beneke and M. Neubert, Nucl. Phys. B **675** (2003) 333 [arXiv:hep-ph/0308039].

[4] A. L. Kagan, Phys. Lett. B **601** (2004) 151 [arXiv:hep-ph/0405134]; A. L. Kagan, arXiv:hep-ph/0407076.

[5] M. Beneke, J. Rohrer and D. Yang, Phys. Rev. Lett. **96** (2006) 141801 [arXiv:hep-ph/0512258]; M. Beneke, J. Rohrer and D. Yang, Nucl. Phys. B **774** (2007) 64 [arXiv:hep-ph/0612290].

[6] U. Amaldi, W. de Boer and H. Furstenau, Phys. Lett. B **260** (1991) 447.

[7] G. D'Ambrosio, G. F. Giudice, G. Isidori and A. Strumia, Nucl. Phys. B **645** (2002) 155 [arXiv:hep-ph/0207036].

[8] S. Heinemeyer, X. Miao, S. Su and G. Weiglein, JHEP **0808** (2008) 087 [arXiv:0805.2359 [hep-ph]]; O. Buchmueller *et al.*, Phys. Lett. B **657** (2007) 87 [arXiv:0707.3447 [hep-ph]].

[9] H. P. Nilles, Phys. Lett. B **115** (1982) 193; A. H. Chamseddine, R. L. Arnowitt and P. Nath, Phys. Rev. Lett. **49** (1982) 970; R. Barbieri, S. Ferrara and C. A. Savoy, Phys. Lett. B **119** (1982) 343; N. Ohta, Prog. Theor. Phys. **70** (1983) 542; L. J. Hall, J. D. Lykken and S. Weinberg, Phys. Rev. D **27** (1983) 2359; S. K. Soni and H. A. Weldon, Phys. Lett. B **126** (1983) 215.

[10] G. W. Bennett *et al.* [Muon G-2 Collaboration], Phys. Rev. D **73** (2006) 072003 [arXiv:hep-ex/0602035].

[11] L. J. Hall, R. Rattazzi and U. Sarid, Phys. Rev. D **50** (1994) 7048 [arXiv:hep-ph/9306309]; M. S. Carena, M. Olechowski, S. Pokorski and C. E. M. Wagner, Nucl. Phys. B **426** (1994) 269 [arXiv:hep-ph/9402253].

[12] T. Blazek, S. Raby and S. Pokorski, Phys. Rev. D **52** (1995) 4151 [arXiv:hep-ph/9504364].

[13] M. S. Carena, S. Mrenna and C. E. M. Wagner, Phys. Rev. D **60** (1999) 075010 [arXiv:hep-ph/9808312].

[14] M. S. Carena, D. Garcia, U. Nierste and C. E. M. Wagner, Nucl. Phys. B **577** (2000) 88 [arXiv:hep-ph/9912516].

[15] G. Isidori and P. Paradisi, Phys. Lett. B **639** (2006) 499 [arXiv:hep-ph/0605012].

[16] U. Nierste, S. Trine and S. Westhoff, Phys. Rev. D **78** (2008) 015006 [arXiv:0801.4938 []]; J. F. Kamenik and F. Mescia, Phys. Rev. D **78** (2008) 014003 [arXiv:0802.3790 []].

[17] C. Hamzaoui, M. Pospelov and M. Toharia, Phys. Rev. D **59** (1999) 095005 [arXiv:hep-ph/9807350];

[18] K. S. Babu and C. F. Kolda, Phys. Rev. Lett. **84** (2000) 228 [arXiv:hep-ph/9909476].

[19] V. M. Abazov *et al.* [D0 Collaboration], Phys. Rev. Lett. **94** (2005) 071802 [arXiv:hep-ex/0410039]; T. Aaltonen *et al.* [CDF Collaboration], Phys. Rev. Lett. **100** (2008) 101802 [arXiv:0712.1708 [hep-ex]].

[20] A. J. Buras, P. H. Chankowski, J. Rosiek and L. Slawianowska, Nucl. Phys. B **619** (2001) 434 [arXiv:hep-ph/0107048]; A. J. Buras, P. H. Chankowski, J. Rosiek and L. Slawianowska, Phys. Lett. B **546** (2002) 96 [arXiv:hep-ph/0207241].

[21] G. Degrassi, P. Gambino and G. F. Giudice, JHEP **0012** (2000) 009 [arXiv:hep-ph/0009337].

[22] M. S. Carena, D. Garcia, U. Nierste and C. E. M. Wagner, Phys. Lett. B **499** (2001) 141 [arXiv:hep-ph/0010003].

[23] A. J. Buras, P. H. Chankowski, J. Rosiek and L. Slawianowska, Nucl. Phys. B **659** (2003) 3 [arXiv:hep-ph/0210145].

[24] J. R. Ellis, J. S. Lee and A. Pilaftsis, Phys. Rev. D **76** (2007) 115011 [arXiv:0708.2079 [hep-ph]].

[25] S. Marchetti, S. Mertens, U. Nierste and D. Stockinger, Phys. Rev. D **79** (2009) 013010 [arXiv:0808.1530 [hep-ph]].

[26] M. Beneke, P. Ruiz-Femenia and M. Spinrath, JHEP **0901**, 031 (2009) [arXiv:0810.3768 [hep-ph]].

[27] L. Hofer, Diplomarbeit (2007).

[28] J. Kublbeck, M. Bohm and A. Denner, Comput. Phys. Commun. **60** (1990) 165; T. Hahn, Comput. Phys. Commun. **140** (2001) 418 [arXiv:hep-ph/0012260].

[29] A. J. Buras and R. Fleischer, Eur. Phys. J. C **11** (1999) 93 [arXiv:hep-ph/9810260].

[30] A. J. Buras, R. Fleischer, S. Recksiegel and F. Schwab, Eur. Phys. J. C **32** (2003) 45 [arXiv:hep-ph/0309012]; A. J. Buras, R. Fleischer, S. Recksiegel and F. Schwab, Phys. Rev. Lett. **92** (2004) 101804 [arXiv:hep-ph/0312259];

[31] T. Yoshikawa, Phys. Rev. D **68** (2003) 054023 [arXiv:hep-ph/0306147]; S. Mishima and T. Yoshikawa, Phys. Rev. D **70** (2004) 094024 [arXiv:hep-ph/0408090].

[32] A. J. Buras, R. Fleischer, S. Recksiegel and F. Schwab, Nucl. Phys. B **697** (2004) 133 [arXiv:hep-ph/0402112]; A. J. Buras, R. Fleischer, S. Recksiegel and F. Schwab, Acta Phys. Polon. B **36** (2005) 2015 [arXiv:hep-ph/0410407]; A. J. Buras, R. Fleischer, S. Recksiegel and F. Schwab, Eur. Phys. J. C **45** (2006) 701 [arXiv:hep-ph/0512032].

[33] C. W. Chiang, M. Gronau, J. L. Rosner and D. A. Suprun, Phys. Rev. D **70** (2004) 034020 [arXiv:hep-ph/0404073]; X. G. He and B. H. J. McKellar, arXiv:hep-ph/0410098.

[34] S. Nandi and A. Kundu, arXiv:hep-ph/0407061; Y. L. Wu and Y. F. Zhou, Phys. Rev. D **71** (2005) 021701 [arXiv:hep-ph/0409221]; Y. L. Wu and Y. F. Zhou, Phys. Rev. D **72** (2005) 034037 [arXiv:hep-ph/0503077].

[35] S. Baek, P. Hamel, D. London, A. Datta and D. A. Suprun, Phys. Rev. D **71** (2005) 057502 [arXiv:hep-ph/0412086]; C. S. Kim, S. Oh and C. Yu, Phys. Rev. D **72** (2005) 074005 [arXiv:hep-ph/0505060]; C. S. Kim, S. Oh and Y. W. Yoon, Phys. Lett. B **665** (2008) 231 [arXiv:0707.2967 [hep-ph]].

[36] Y. Y. Keum, H. n. Li and A. I. Sanda, Phys. Lett. B **504** (2001) 6 [arXiv:hep-ph/0004004]; Y. Y. Keum, H. N. Li and A. I. Sanda, Phys. Rev. D **63** (2001) 054008 [arXiv:hep-ph/0004173].

[37] X. q. Li and Y. d. Yang, Phys. Rev. D **72** (2005) 074007 [arXiv:hep-ph/0508079].

[38] C. W. Bauer, D. Pirjol, I. Z. Rothstein and I. W. Stewart, Phys. Rev. D **70** (2004) 054015 [arXiv:hep-ph/0401188]; C. W. Bauer, I. Z. Rothstein and I. W. Stewart, Phys. Rev. D **74** (2006) 034010 [arXiv:hep-ph/0510241]; A. R. Williamson and J. Zupan, Phys. Rev. D **74** (2006) 014003 [Erratum-ibid. D **74** (2006) 03901] [arXiv:hep-ph/0601214].

[39] R. Fleischer, S. Recksiegel and F. Schwab, Eur. Phys. J. C **51** (2007) 55 [arXiv:hep-ph/0702275]; R. Fleischer, S. Jager, D. Pirjol and J. Zupan, Phys. Rev. D **78** (2008) 111501 [arXiv:0806.2900 [hep-ph]].

[40] S. Baek and D. London, Phys. Lett. B **653** (2007) 249 [arXiv:hep-ph/0701181]; S. Baek, C. W. Chiang and D. London, Phys. Lett. B **675** (2009) 59 [arXiv:0903.3086 [hep-ph]].

[41] M. Gronau and J. L. Rosner, Phys. Rev. D **59** (1999) 113002 [arXiv:hep-ph/9809384].

[42] E. Barberio *et al.* [Heavy Flavor Averaging Group], arXiv:0808.1297 [hep-ex], http://www.slac.stanford.edu/xorg/hfag/.

[43] Y. Grossman, M. Neubert and A. L. Kagan, JHEP **9910** (1999) 029 [arXiv:hep-ph/9909297].

[44] G. Buchalla, G. Hiller and G. Isidori, Phys. Rev. D **63** (2000) 014015 [arXiv:hep-ph/0006136].

[45] V. Barger, C. W. Chiang, P. Langacker and H. S. Lee, Phys. Lett. B **598** (2004) 218 [arXiv:hep-ph/0406126].

[46] W. S. Hou, M. Nagashima and A. Soddu, Phys. Rev. Lett. **95** (2005) 141601 [arXiv:hep-ph/0503072].

[47] S. Khalil, Phys. Rev. D **72** (2005) 035007 [arXiv:hep-ph/0505151].

[48] R. Fleischer, Phys. Lett. B **332** (1994) 419.

[49] M. Beneke, X. Q. Li and L. Vernazza, Eur. Phys. J. C **61** (2009) 429 [arXiv:0901.4841 [hep-ph]].

[50] A. J. Buras, M. Jamin, M. E. Lautenbacher and P. H. Weisz, Nucl. Phys. B **370** (1992) 69 [Addendum-ibid. B **375** (1992) 501]; A. J. Buras, M. Jamin, M. E. Lautenbacher and P. H. Weisz, Nucl. Phys. B **400** (1993) 37 [arXiv:hep-ph/9211304]; A. J. Buras, M. Jamin and M. E. Lautenbacher, Nucl. Phys. B **400** (1993) 75 [arXiv:hep-ph/9211321]; M. Ciuchini, E. Franco, G. Martinelli and L. Reina, Phys. Lett. B **301** (1993) 263 [arXiv:hep-ph/9212203]; M. Ciuchini, E. Franco, G. Martinelli and L. Reina, Nucl. Phys. B **415** (1994) 403 [arXiv:hep-ph/9304257].

[51] A. J. Buras, M. Misiak, M. Munz and S. Pokorski, Nucl. Phys. B **424** (1994) 374 [arXiv:hep-ph/9311345].

[52] J.D. Bjorken, *Nucl. Phys. Proc. Suppl.* **B 11** (1989) 325.

[53] C. Amsler *et al.* [Particle Data Group], Phys. Lett. B **667** (2008) 1.

[54] M. Beneke and S. Jager, Nucl. Phys. B **751** (2006) 160 [arXiv:hep-ph/0512351]; M. Beneke and S. Jager, Nucl. Phys. B **768** (2007) 51 [arXiv:hep-ph/0610322]; M. Bartsch, G. Buchalla and C. Kraus, arXiv:0810.0249 [hep-ph]; M. Beneke, T. Huber and X. Q. Li, Nucl. Phys. B **832** (2010) 109 [arXiv:0911.3655 [hep-ph]].

[55] S. P. Martin, arXiv:hep-ph/9709356; M. Drees, arXiv:hep-ph/9611409.

[56] A. J. Buras, P. Gambino, M. Gorbahn, S. Jager and L. Silvestrini, Phys. Lett. B **500** (2001) 161 [arXiv:hep-ph/0007085]; M. Blanke, A. J. Buras, D. Guadagnoli and C. Tarantino, JHEP **0610** (2006) 003 [arXiv:hep-ph/0604057].

[57] L. Girardello and M. T. Grisaru, Nucl. Phys. B **194** (1982) 65.

[58] J. Berger and Y. Grossman, Phys. Lett. B **675** (2009) 365 [arXiv:0811.1019 [hep-ph]].

[59] G. Colangelo, E. Nikolidakis and C. Smith, Eur. Phys. J. C **59**, 75 (2009) [arXiv:0807.0801 [hep-ph]]; C. Smith, arXiv:0909.4444 [hep-ph].

[60] H. Baer, M. Brhlik, D. Castano and X. Tata, Phys. Rev. D **58** (1998) 015007 [arXiv:hep-ph/9712305]; B. Dudley and C. Kolda, Phys. Rev. D **79** (2009) 015011 [arXiv:0805.4565 [hep-ph]].

[61] P. Z. Skands *et al.*, JHEP **0407** (2004) 036 [arXiv:hep-ph/0311123].

[62] M. Gorbahn, S. Jager, U. Nierste and S. Trine, arXiv:0901.2065 [hep-ph].

[63] T. Banks, Nucl. Phys. B **303** (1988) 172.

[64] T. Kinoshita, J. Math. Phys. **3** (1962) 650; T. D. Lee and M. Nauenberg, Phys. Rev. **133** (1964) B1549.

[65] D. M. Pierce, J. A. Bagger, K. T. Matchev and R. j. Zhang, Nucl. Phys. B **491**, 3 (1997) [arXiv:hep-ph/9606211].

[66] H. E. Logan and U. Nierste, Nucl. Phys. B **586** (2000) 39 [arXiv:hep-ph/0004139].

[67] A. Denner and T. Sack, Nucl. Phys. B **347** (1990) 203.

[68] P. Gambino, P. A. Grassi and F. Madricardo, Phys. Lett. B **454** (1999) 98 [arXiv:hep-ph/9811470].

[69] J. A. Casas, A. Lleyda and C. Munoz, Nucl. Phys. B **471** (1996) 3 [arXiv:hep-ph/9507294]; J. A. Casas and S. Dimopoulos, Phys. Lett. B **387** (1996) 107 [arXiv:hep-ph/9606237].

[70] S. Pokorski, J. Rosiek and C. A. Savoy, Nucl. Phys. B **570** (2000) 81 [arXiv:hep-ph/9906206].

[71] G. Degrassi, P. Gambino and P. Slavich, Phys. Lett. B **635** (2006) 335 [arXiv:hep-ph/0601135]; G. Degrassi, P. Gambino and P. Slavich, Comput. Phys. Commun. **179** (2008) 759 [arXiv:0712.3265 [hep-ph]].

[72] M. Misiak *et al.*, Phys. Rev. Lett. **98** (2007) 022002 [arXiv:hep-ph/0609232]; M. Misiak and M. Steinhauser, Nucl. Phys. B **764** (2007) 62 [arXiv:hep-ph/0609241].

[73] A. L. Kagan and M. Neubert, Eur. Phys. J. C **7** (1999) 5 [arXiv:hep-ph/9805303].

[74] G. Buchalla, G. Hiller, Y. Nir and G. Raz, JHEP **0509** (2005) 074 [arXiv:hep-ph/0503151].

[75] W. Altmannshofer, A. J. Buras and P. Paradisi, Phys. Lett. B **669** (2008) 239 [arXiv:0808.0707 [hep-ph]].

[76] Y. Nir and H. R. Quinn, Phys. Rev. Lett. **67** (1991) 541; M. Gronau, Phys. Lett. B **265** (1991) 389.

[77] M. Gronau, O. F. Hernandez, D. London and J. L. Rosner, Phys. Rev. D **50** (1994) 4529 [arXiv:hep-ph/9404283]; M. Gronau, O. F. Hernandez, D. London and J. L. Rosner, Phys. Rev. D **52** (1995) 6374 [arXiv:hep-ph/9504327].

[78] H. J. Lipkin, Phys. Lett. B **445** (1999) 403 [arXiv:hep-ph/9810351].

[79] M. Gronau and J. L. Rosner, Phys. Rev. D **74** (2006) 057503 [arXiv:hep-ph/0608040]; M. Gronau and J. L. Rosner, Phys. Lett. B **644** (2007) 237 [arXiv:hep-ph/0610227]; S. Baek, C. W. Chiang, M. Gronau, D. London and J. L. Rosner, Phys. Lett. B **678** (2009) 97 [arXiv:0905.1495 [hep-ph]].

[80] G. Zweig, CERN Report No. 8419 TH 412, 1964 (unpublished); reprinted in Developments in the Quark Theory of Hadrons, edited by D. B. Lichtenberg and S. P. Rosen (Hadronic Press, Massachusetts, 1980); S. Okubo, Phys. Lett. 5, 165 (1963); Phys. Rev. D 16, 2336 (1977); J. Iizuka, K. Okada, and O. Shito, Prog. Th. Phys. 35, 1061 (1966); J. Iizuka, Prog. Th. Phys. Suppl. 37-38, 21 (1966).

[81] T. Feldmann, M. Jung and T. Mannel, JHEP **0808** (2008) 066 [arXiv:0803.3729 [hep-ph]].

[82] A. Hocker, H. Lacker, S. Laplace and F. Le Diberder, Eur. Phys. J. C **21** (2001) 225 [arXiv:hep-ph/0104062].

[83] G. Buchalla and A. J. Buras, Nucl. Phys. B **400** (1993) 225.

[84] A. J. Buras and M. Münz, Phys. Rev. D **52** (1995) 186 [arXiv:hep-ph/9501281].

[85] G. Buchalla, A. J. Buras and M. E. Lautenbacher, Rev. Mod. Phys. **68** (1996) 1125 [arXiv:hep-ph/9512380].

[86] B. Aubert *et al.* [BABAR Collaboration], Phys. Rev. Lett. **93** (2004) 081802 [arXiv:hep-ex/0404006]; M. Iwasaki *et al.* [Belle Collaboration], Phys. Rev. D **72** (2005) 092005 [arXiv:hep-ex/0503044].

[87] C. Bobeth, G. Hiller and G. Piranishvili, JHEP **0807** (2008) 106 [arXiv:0805.2525 [hep-ph]].

[88] U. Egede, T. Hurth, J. Matias, M. Ramon and W. Reece, JHEP **0811** (2008) 032 [arXiv:0807.2589 [hep-ph]]. W. Altmannshofer, P. Ball, A. Bharucha, A. J. Buras, D. M. Straub and M. Wick, JHEP **0901** (2009) 019 [arXiv:0811.1214 [hep-ph]].

[89] V. Lubicz and C. Tarantino, Nuovo Cim. **123B** (2008) 674 [arXiv:0807.4605 [hep-lat]].

[90] A. Abulencia *et al.* [CDF Collaboration], Phys. Rev. Lett. **97** (2006) 242003 [arXiv:hep-ex/0609040].

[91] A. Lenz *et al.*, arXiv:1008.1593 [hep-ph].

[92] P. Langacker and D. London, Phys. Rev. D **38** (1988) 886; E. Nardi, E. Roulet and D. Tommasini, Nucl. Phys. B **386** (1992) 239.

[93] W. Altmannshofer, A. J. Buras, D. M. Straub and M. Wick, JHEP **0904** (2009) 022 [arXiv:0902.0160 [hep-ph]].

[94] P. Langacker, Rev. Mod. Phys. **81** (2009) 1199 [arXiv:0801.1345 [hep-ph]].

[95] V. Barger, L. Everett, J. Jiang, P. Langacker, T. Liu and C. Wagner, Phys. Rev. D **80** (2009) 055008 [arXiv:0902.4507 [hep-ph]]; Q. Chang, X. Q. Li and Y. D. Yang, JHEP **0905** (2009) 056 [arXiv:0903.0275 [hep-ph]];

[96] T. G. Rizzo, Phys. Rev. D **59** (1999) 015020 [arXiv:hep-ph/9806397].

DANKSAGUNG

An dieser Stelle möchte ich mich bei all jenen bedanken, die zum Gelingen dieser Arbeit beigetragen haben.

Mein erster Dank geht an Prof. Ulrich Nierste für das interessante und vielseitige Thema, sowie für die Freiheit, die er mir bei den Forschungsprojekten gewährt hat. Seine freundschaftliche Betreuung und seine in jeder Hinsicht uneingeschränkte Unterstützung kann nicht genug gewürdigt werden.

Bei Prof. Dieter Zeppenfeld bedanke ich mich für die bereitwillige Übernahme des Korreferats.

Des Weiteren danke ich Dominik Scherer und Leonardo Vernazza für die angenehme und ergiebige Zusammenarbeit an unseren gemeinsamen Projekten in Ref. [1, 2]. Ferner möchte ich mich bedanken bei Andreas Crivellin für die vielen an(st)re(n)genden Diskussionen, bei denen ich häufig einen neuen Blickwinkel und viele Erkenntnisse gewonnen habe. Peter Marquard bin ich dankbar dafür, dass er bei Computerproblemen stets als Ansprechpartner zur Verfügung stand. Bei meinem Bürokollegen Nikolai Zerf sowie auch bei allen anderen Mitgliedern des Instituts bedanke ich mich für die angenehme Arbeitsatmosphäre.

Für das Auffinden unzähliger Tipp-, Wort- und Grammatikfehler danke ich Daximilian Dachsmüller, Holger Drees, Simon Fink, Martin Hofer, Wiebke Hofer, Jens Ilg, Alexander Krätke, Melanie Kullmer, Roland Kuschill, Sabine Paarmann, Paul Ruhland, Christoph Scheben, Nina Schmalzridt und Markus Völker.

Die Arbeit wurde gefördert durch das Graduiertenkolleg "Hochenergie- und Teilchenastrophysik" der DFG, sowie vor allem durch das Evangelische Studienwerk Villigst.

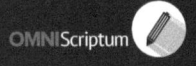

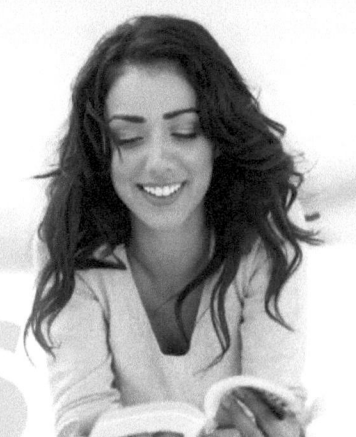

MIX
Papier aus verantwortungsvollen Quellen
Paper from responsible sources
FSC® C105338

Printed by Books on Demand GmbH, Norderstedt / Germany